T0189965

Generalizations of Fuzzy Information Measures

Anshu Ohlan · Ramphul Ohlan

Generalizations of Fuzzy Information Measures

Anshu Ohlan
Department of Mathematics
Deenbandhu Chhotu Ram University of Science and Technology
Murthal, Haryana
India

Ramphul Ohlan
Institute of Management Studies and Research
Maharshi Dayanand University
Rohtak, Haryana
India

ISBN 978-3-319-83401-6 ISBN 978-3-319-45928-8 (eBook)
DOI 10.1007/978-3-319-45928-8

Printed on acid-free paper

This Springer imprint is published by Springer Nature
The registered company is Springer International Publishing AG
The registered company address is: Gewerbestrasse 11, 6330 Cham, Switzerland

Foreword

The novel generalizations of information and divergence measures have deep applications in the natural and social sciences. The focus of this book is on applications of development in generalized measures of information and divergence in fuzzy and intuitionistic fuzzy environment to solve the problems related to the multi-criteria decision-making, linguistic variables, pattern recognition and medical diagnosis. The generalizations of information measures are fresh and well introduced; the numerical examples are quite convincing.

The new generalized *R*-norm fuzzy information and divergence measures are introduced and validated by characterization and generalization of various measures of fuzzy information. It is followed by establishing relations between generalized fuzzy entropy measure and their fuzzy divergence measures, which improved their simplicity, consistency and flexibility. The comparative applications of parametric generalized exponential fuzzy divergence measure are provided in the context of strategic decision-making. In addition, a sequence of fuzzy mean difference divergence measures is introduced with a number of inequalities among them. In doing so, their efficiency is achieved in pattern recognition and compound linguistic variables. A newly generalized fuzzy divergence measure and its application to multi-criteria decision-making and pattern recognition are given. This book also introduces a generalized fuzzy Hellinger's divergence measure, which is followed by providing illustration of its application in the field of medical diagnosis. Finally, the study of a novel and efficient exponential divergence measure in intuitionistic fuzzy environment is provided with its application in multi-attribute decision-making.

To conclude, I strongly state that the book presents a very interesting applied research contribution of innovative mathematical methods to address problems arising in modern research in sciences.

Brazil

I.J. Taneja

Contents

Chapter 1
Fundamentals of Fuzzy Information Measures

Information theory as a branch of mathematics is novel that worldwide deals with a variety of fields. The term 'Information' is concerned with our routine life, like energy, but it is difficult to define in view of the fact that it has no component parts, it can only be measured. The study of information theory was first carried out by Nyquist [63, 64] and latter by Hartley [32]. However, as a precise mathematical concept information is the product of the elementary paper '*The Mathematical Theory of Communication*' of Shannon [74]. Indeed, Shannon was first to use the word 'Entropy' to measure an uncertain degree of the randomness in a probability distribution. Shannon developed a quantitative measure of uncertainty (information) in the logarithmic nature called entropy. This concept of entropy has been applied widely in different areas such as communication theory, information processing, fuzzy theory, pattern recognition, image segmentation, decision-making, medical diagnosis, etc.

1.1 Introduction to Fundamental Concepts of Fuzzy Set Theory

Fuzzy set theory was first proposed by Zadeh [94] as a generalization of classical (crisp) set theory. It is based on the idea that in the real world a precise description of many situations is nearly impossible, and that imprecisely defined classes play an important role in human thinking and natural language. In natural language, many words have ambiguous meanings that are properly quantified and analyzed with the help of fuzzy set theory. Over the last six decades, research on fuzzy set theory and its application to different areas have been extended evenly. The theory of fuzzy sets has been proven useful in the context of decision theory, pattern recognition, medical diagnosis, control theory, etc.

A. Ohlan and R. Ohlan, *Generalizations of Fuzzy Information Measures*,
DOI 10.1007/978-3-319-45928-8_1

Probability theory provides the numerical information based on measurement, whereas fuzzy theory provides the linguistic information based on perception. For example, '*Sita is 27*', a numerical information but '*Sita is young*' is linguistic information. The fuzzy set theory has resulted in an important breakthrough in the modelling of the vagueness, imprecision and uncertainty that characterize human knowledge [96].

Fuzziness, a feature of uncertainty, is found in our judgment, in our words and the way in which we process information. According to Kosko [49] there is conceptual and theoretical difference between randomness and fuzziness. However, randomness describes the uncertainty of event occurrence whereas fuzziness describes event ambiguity. It measures the degree to which an event occurs, not whether it occurs. Zadeh [97] discussed on the concept of probability theory and fuzzy logic as complimentary rather than competitive. According to Félix [28] 'Fuzzy sets are a natural and intuitively plausible way to represent and handle expressions such as '*much later*', '*rises slowly*' or '*more or less* 15 *degrees*', and may be thought of as a generalization of the representation of qualitative and quantitative information.'

1.1.1 Crisp (Classical) Set

In our life we mostly use the crisp sets. A crisp set is a collection of things that belongs to a definition. In a crisp set any item either belongs or does not belong to the set. For a crisp set A, the characteristic function is defined as $\chi_A : X \rightarrow \{0, 1\}$

$$\chi_A = \begin{cases} 1, & \text{if } x \in A \\ 0, & \text{if } x \notin A \end{cases}$$

The set $A = \{(x, \chi_A(x))\}$ is said to be a crisp set, i.e. the characteristic function having values either of 0 or 1 in the classical set. Thus, the crisp set is also known as the classical (or well-defined sharp) set. Finally, a classical set A is the collection of all those members of universe of discourse X for which $\chi_A(x) = 1$.

1.1.2 Fuzzy Set

Formally, a fuzzy set A defined on a universe of discourse $X = \{x_1, x_2, \ldots, x_n\}$ is given by Zadeh [94]:

$$A = \{\langle x, \mu_A(x)\rangle / x \in X\},$$

where $\mu_A : X \rightarrow [0, 1]$ is the membership function of A. The membership value $\mu_A(x)$ describes the degree of the belongingness of $x \in X$ in A. The closer the value

Table 1.1 Numerical values of membership function $Y(x)$

Person	Age (x)	Membership value $Y(x)$
A	20	1.00
B	26	0.90
C	30	0.50
D	34	0.10
E	36	0.00

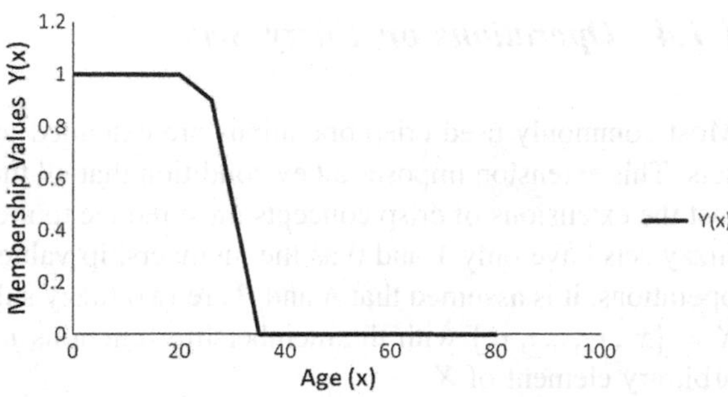

Fig. 1.1 Graph of the membership values $Y(x)$

of $\mu_A(x)$ is to 1, the more x belongs to A. When $\mu_A(x)$ is valued in $\{0, 1\}$, it is the characteristic function of a crisp (i.e. non-fuzzy) set. The characteristic function of a fuzzy set is called membership function, the role of which has properly been explained by Singpurwalla and Booker [78] in probability measures of fuzzy sets.

The following example explains the concept of fuzzy set.

Example 1.1 The concept of fuzzy set can be explained with one of the example of 'youthfulness' of different aged people. Let us assume X (the universe of discourse) is the set of people. A fuzzy subset young, which answer the question 'to which degree the person x is young?' is denoted by $Y(x)$ is defined as

$$Y(x) = \left\{ \begin{array}{cc} 1, & \text{if age}(x) \leq 25 \\ (35 - \text{age}(x))/10, & \text{if } 25 < \text{age}(x) \leq 35 \\ 0, & \text{if age}(x) > 35 \end{array} \right\}.$$

Table 1.1 represents the degree of youth of different aged persons.

The above example is explained more precisely in Fig. 1.1.

1.1.3 Difference Between Crisp Set and Fuzzy Set

The key difference between a crisp set and a fuzzy set is their membership function. A crisp set has unique membership function, whereas a fuzzy set can have an infinite number of membership functions to represent it. For example, we can define a possible membership function for the set of real numbers close to 0 as follows [48]:

$$\mu_A(x) = \frac{1}{1+10x^2}; \quad x \in R.$$

Here the number 3 is assigned a grade of 0.01, the number 1 is assigned a grade of 0.09 and the number 0 is assigned a grade of 1. Thus, the concepts of crisp sets are incorporated by fuzzy sets as a special case.

1.1.4 Operations on Fuzzy Sets

Most commonly used crisp operations are extended up by the operations on fuzzy sets. This extension imposes a key condition that all the fuzzy operations, which are just the extensions of crisp concepts must reduce to their natural meaning where the fuzzy sets have only 1 and 0 as the membership values. For defining the following operations, it is assumed that A and B are two fuzzy subsets of universe of discourse $X = \{x_1, x_2, \ldots, x_n\}$ with the membership functions $\mu_A(x)$ and $\mu_B(x)$; x denotes an arbitrary element of X.

Zadeh [94] gave some notions related to fuzzy sets which are as follows:

(1) **Containment**; $A \subset B \Leftrightarrow \mu_A(x) \le \mu_B(x)$ for all $x \in X$.
(2) **Equality**; $A = B \Leftrightarrow \mu_A(x) = \mu_B(x)$ for all $x \in X$.
(3) **Complement**; $\overline{A}$ = Complement of $A \Leftrightarrow \mu_{\overline{A}}(x) = 1 - \mu_A(x)$ for all $x \in X$.
(4) **Union**; $A \cup B$ = Union of A and $B \Leftrightarrow \mu_{A\cup B}(x) = \max\{\mu_A(x), \mu_B(x)\}$ for all $x \in X$.
(5) **Intersection**; $A \cap B$ = Intersection of A and $B \Leftrightarrow \mu_{A\cap B}(x) = \min\{\mu_A(x), \mu_B(x)\}$ for all $x \in X$.
(6) **Product**; AB = Product of A and $B \Leftrightarrow \mu_{AB}(x) = \mu_A(x)\mu_B(x)$ for all $x \in X$.
(7) **Sum**; $A + B$ = Sum of A and $B \Leftrightarrow \mu_{A\oplus B}(x) = \mu_A(x) + \mu_B(x) - \mu_A(x)\mu_B(x)$ for all $x \in X$.

1.1.5 Fundamental Properties of Fuzzy Sets

The fundamental properties of fuzzy sets are as follows:

(i) Idempotent law:

$$A \cap A = A; \quad A \cup A = A; \quad \overline{\overline{A}} = A.$$

(ii) Commutative law:

$$A \cap B = B \cap A; \quad A \cup B = B \cup A.$$

(iii) Associative law:

$$(A \cap B) \cap C = A \cap (B \cap C); \quad (A \cup B) \cup C = A \cup (B \cup C).$$

(iv) Distributive law:

$$(A \cap B) \cup C = (A \cup C) \cap (B \cup C); \quad (A \cup B) \cap C = (A \cap C) \cup (B \cap C).$$

(v) Absorption law:

$$A \cup (A \cap B) = A; \quad A \cap (A \cup B) = A.$$

(vi) Zero Law:

$$A \cup U = U; \quad A \cap \phi = \phi.$$

(vii) Identity Law:

$$A \cup \phi = A; \quad A \cap U = A.$$

(viii) De Morgan's law:

$$\overline{A \cup B} = \overline{A} \cap \overline{B}; \quad \overline{A \cap B} = \overline{A} \cup \overline{B}.$$

1.1.6 Some Basic Concepts

Some important basic concepts related to fuzzy set theory are defined as given below

(i) **Equality of Two Fuzzy Sets**

Two fuzzy sets are said to be equal if and only if $\mu_A(x_i) = \mu_B(x_i)$; $\forall x_i \in X$.

(ii) **Standard Fuzzy Sets**

Fuzzy sets are said to be standard if $\mu_A(x_i) \leq 0.5$; $\forall x_i \in X$.

(iii) **Support of a Fuzzy Set**

Given a fuzzy set A which is a subset of the universal set $X = \{x_1, x_2, \ldots, x_n\}$, the support of A, denoted by Supp(A), is an ordinary set defined as the set of elements whose degree of membership in A is greater than 0, i.e. $\text{Supp}(A) = \{x_i \in X / \mu_A(x_i) > 0\}$.

1.2 Fuzzy Entropy Measures and Their Generalizations

There have been several attempts to quantify the uncertainty associated with fuzzy sets. Zadeh [95] first introduced the concept of fuzzy entropy as a measure of fuzziness. Fuzzy entropy is an important concept for measuring fuzzy information. A measure of the fuzzy entropy of a fuzzy set is a measure of the fuzziness of the set. Fuzzy entropy have been widely applied in many fields such as pattern recognition, image processing, speech recognition, bioinformatics, fuzzy aircraft control, feature selection, etc.

Since $\mu_A(x_i)$ and $1-\mu_A(x_i)$ for all $i=1,2,\ldots,n$ give the same degree of fuzziness, therefore, De Luca and Termini [21] introduced an axiomatic structure of the measure of fuzzy entropy corresponding to Shannon [74] entropy as

$$H(A) = -\sum_{i=1}^{n}\left[\mu_A(\mathrm{x_i})\log\mu_A(\mathrm{x_i}) + (1-\mu_A(\mathrm{x_i}))\log(1-\mu_A(\mathrm{x_i}))\right] \tag{1.1}$$

A measure of fuzziness $H(A)$ of a fuzzy set A should satisfy at least the following four properties (P1–P4):

(P1) $H(A)$ is minimum if and only if A is a crisp set, i.e.

$$\mu_A(x_i) = 0 \text{ or } 1 \quad \text{for all } x_i\text{: } i = 1,2,\ldots n.$$

(P2) $H(A)$ is maximum if and only if A is most fuzzy set, i.e.

$$\mu_A(x_i) = 0.5 \quad \text{for all } x_i\text{: } i = 1,2,\ldots n.$$

(P3) $H(A) \geq H(A^*)$, where A^* is sharpened version of A.
where a fuzzy set A^* is the sharpened version of A if the following conditions are satisfied:

$$\begin{aligned} \mu_{A^*}(x_i) &\leq \mu_A(x_i), \quad \text{if } \mu_A(x_i) \leq 0.5;\ \forall i \\ \text{and } \mu_{A^*}(x_i) &\geq \mu_A(x_i), \quad \text{if } \mu_A(x_i) \geq 0.5;\ \forall i. \end{aligned}$$

(P4) $H(A) = H(\overline{A})$, where $\overline{A}$ is the complement of A.

It may be noted that if $x_1, x_2, \ldots, x_n$ are members of the universe of discourse, then all $\mu_A(x_1), \mu_A(x_2), \mu_A(x_3), \ldots, \mu_A(x_n)$ lie between 0 and 1, but these are not the probabilities because their sum is not unity. However,

$$\Phi_A(x_i) = \frac{\mu_A(x_i)}{\sum_{i=1}^{n}\mu_A(x_i)}; \quad i = 1,2,\ldots n \text{ is a probability distribution.}$$

Kaufmann [47] defined entropy of a fuzzy set A having n support points by

$$H(A) = -\frac{1}{\log n}\sum_{i=1}^{n} \Phi_A(x_i)\log \Phi_A(x_i) \tag{1.2}$$

Bhandari and Pal [8] extended the probabilistic exponential entropy idea of Pal and Pal [66] to the fuzzy phenomenon and defined the exponential fuzzy entropy for a fuzzy set A as

$$H_e(A) = \frac{1}{n(\sqrt{e}-1)}\sum_{i=1}^{n} [\mu_A(x_i)e^{(1-\mu_A(x_i))} + (1-\mu_A(x_i))e^{\mu_A(x_i)} - 1] \tag{1.3}$$

Kapur [46] argues that fuzzy entropy measures uncertainty due to fuzziness of information, while probabilistic entropy measures uncertainty due to the information being available in terms of a probability distribution only. The advantage of making probability theory work in fuzzy environment is to deal with a different kind of uncertainties that may arise within the same problem.

1.3 Measures of Fuzzy Divergence

In fuzzy context, several measures have been proposed in order to measure the degree of difference between two fuzzy sets. Measure of fuzzy divergence between two fuzzy sets gives the difference between two fuzzy sets and this measure of difference between two fuzzy sets is called the fuzzy divergence measure.

Analogous to Kullback and Leibler [51] measure of divergence, Bhandari and Pal [8] introduced divergence measure between two fuzzy sets A and B of universe of discourse $X = \{x_1, x_2, \ldots, x_n\}$ having the membership values $\mu_A(x_i)$ and $\mu_B(x_i)$, as

$$I(A:B) = \sum_{i=1}^{n}\left[\mu_A(x_i)\log\frac{\mu_A(x_i)}{\mu_B(x_i)} + (1-\mu_A(x_i))\log\frac{1-\mu_A(x_i)}{1-\mu_B(x_i)}\right] \tag{1.4}$$

satisfying the conditions

(i) $I(A:B) \geq 0$
(ii) $I(A:B) = 0$ if $A = B$
(iii) $I(A:B)$ is a convex function of A and B.

Bhandari and Pal [8] also defined the fuzzy symmetric divergence measure

$$J(A:B) = I(A:B) + I(B:A) = \sum_{i=1}^{n}\left[(\mu_A(x_i) - \mu_B(x_i))\log\frac{\mu_A(x_i)(1-\mu_B(x_i))}{\mu_B(x_i)(1-\mu_A(x_i))}\right] \tag{1.5}$$

A general study of the axiomatic definition of a divergence measure for fuzzy sets was also presented in Bouchon-Meunier et al. [10]. As a significant content in fuzzy mathematics, the research on divergence measures between fuzzy sets has received more attention. In recent years, some definitions of generalized measures of fuzzy divergence have been proposed.

Afterwards, Shang and Jiang [73] provided a modified version of the fuzzy divergence measure of Bhandari and Pal [8] with the novel idea of Lin [56] and defined as

$$D(A:B)=\frac{1}{n}\sum_{i=1}^{n}\left[\mu_A(x_i)\log\frac{\mu_A(x_i)}{\left(\frac{\mu_A(x_i)+\mu_B(x_i)}{2}\right)}+(1-\mu_A(x_i))\log\frac{1-\mu_A(x_i)}{\left(\frac{2-\mu_A(x_i)-\mu_B(x_i)}{2}\right)}\right] \tag{1.6}$$

Fan and Xie [27] proposed the fuzzy information of discrimination of A against B corresponding to the exponential fuzzy entropy of Pal and Pal [66] and is given by

$$I(A,B)=\sum_{i=1}^{n}\left[1-(1-\mu_A(x_i))e^{(\mu_A(x_i)-\mu_B(x_i))}-\mu_A(x_i)e^{(\mu_B(x_i)-\mu_A(x_i))}\right] \tag{1.7}$$

Thereafter Couso et al. [20] define that if X is a universe of discourse and $F(X)$ is the set of all fuzzy subsets, a mapping $D:F(X)\times F(X)\rightarrow R$ is a divergence measure between fuzzy subsets if and only if for each $A,B,C\in F(X)$, the following axioms hold:

d_1: $D(A:B)=D(B:A)$
d_2: $D(A:A)=0$
d_3: $\max\{D(A\cup C,B\cup C),D(A\cap C,B\cap C)\}\leq D(A,B)$

Non-negativity of $D(A:B)$ is the natural assumption.

Montes et al. [62] studied the special classes of divergence measures and used the link between fuzzy and probabilistic uncertainty. It also studied widely the divergence measure for fuzzy sets as a particular case.

Hooda [37] presented a fuzzy divergence measure corresponding to Havrada-Charvat [34] measure of divergence which is given by

$$I_\alpha(A,B)=\frac{1}{\alpha-1}\sum_{i=1}^{n}[\mu_A^\alpha(x_i)\mu_B^{1-\alpha}(x_i)+(1-\mu_A(x_i))^\alpha(1-\mu_B(x_i))^{1-\alpha}-1], \quad \alpha\neq1,\ \alpha>0. \tag{1.8}$$

Parkash et al. [68] proposed a fuzzy divergence measure corresponding to Ferreri [29] probabilistic measure of divergence given by

$$\overline{I_\alpha}(A:B) = \frac{1}{a}\sum_{i=1}^{n}\left[(1+a\mu_A(x_i))\log\frac{1+a\mu_A(x_i)}{1+a\mu_B(x_i)} + \{1+a(1-\mu_A(x_i))\}\log\frac{1+a(1-\mu_A(x_i))}{1+a(1-\mu_B(x_i))}\right]; \quad a>0 \tag{1.9}$$

and another one given by Parkash et al. [68] is

$$\begin{aligned} I_a(A:B) = &\sum_{i=1}^{n}\left[\mu_A(x_i)\log\frac{\mu_A(x_i)}{\mu_B(x_i)} + (1-\mu_A(x_i))\log\frac{1-\mu_A(x_i)}{1-\mu_B(x_i)}\right] \\ &-\frac{1}{a}\sum_{i=1}^{n}\left[(1+a\mu_A(x_i))\log\frac{1+a\mu_A(x_i)}{1+a\mu_B(x_i)} + \{1+a(1-\mu_A(x_i))\}\log\frac{1+a(1-\mu_A(x_i))}{1+a(1-\mu_B(x_i))}\right] \end{aligned} \tag{1.10}$$

Corresponding to Renyi [70] and Sharma and Mittal [75] generalized measure of divergence Bajaj and Hooda [5] provided the generalized fuzzy divergence measures which are given by

$$\begin{aligned} D_\alpha(A,B) = \frac{1}{\alpha-1}\sum_{i=1}^{n}\log[\mu_A^\alpha(x_i)\mu_B^{1-\alpha}(x_i) + (1-\mu_A(x_i))^\alpha(1-\mu_B(x_i))^{1-\alpha}], \\ \alpha\neq 1,\ \alpha>0. \end{aligned} \tag{1.11}$$

and

$$\begin{aligned} D_{\alpha,\beta}(A,B) = \frac{1}{2^{1-\beta}-1}\sum_{i=1}^{n}\left[(\mu_A^\alpha(x_i)\mu_B^{1-\alpha}(x_i) + (1-\mu_A(x_i))^\alpha(1-\mu_B(x_i))^{1-\alpha})^{\frac{\beta-1}{\alpha-1}} - 1\right], \\ \alpha\neq 1,\ \alpha>0,\ \beta\neq 1,\ \beta>0. \end{aligned} \tag{1.12}$$

Singh and Tomar [77], Tomar and Ohlan [84] have defined and studied some of symmetric and non-symmetric fuzzy divergence measures analogous to probabilistic divergence measures and inequalities among them. They also presented a number of refinement of inequalities among fuzzy divergence measures. Bhatia and Singh [9] proposed three families of fuzzy divergence

(i) The fuzzy divergence measure corresponding to Taneja [82] Arithmetic-Geometric divergence measure is given by

$$T(A,B) = \sum_{i=1}^{n}\left[\begin{aligned} &\frac{\mu_A(x_i)+\mu_B(x_i)}{2}\log\frac{(\mu_A(x_i)+\mu_B(x_i))}{2\sqrt{\mu_A(x_i)\mu_B(x_i)}} \\ &+\frac{2-\mu_A(x_i)-\mu_B(x_i)}{2}\log\frac{(2-\mu_A(x_i)-\mu_B(x_i))}{2\sqrt{(1-\mu_A(x_i))(1-\mu_B(x_i))}} \end{aligned}\right] \tag{1.13}$$

(ii) A generalized triangular discrimination between two arbitrary fuzzy sets A and B

$$\Delta_\alpha(A,B) = \sum_{i=1}^{n} (\mu_A(x_i) - \mu_B(x_i))^{2\alpha} \left[\frac{1}{(\mu_A(x_i) + \mu_B(x_i))^{2\alpha-1}} + \frac{1}{(2 - \mu_A(x_i) - \mu_B(x_i))^{2\alpha-1}} \right] \tag{1.14}$$

(iii) A (α, β) class of measures of fuzzy divergence for two arbitrary fuzzy sets A and B corresponding to Taneja [82]

$$D_\alpha^\beta(A,B) = \frac{1}{\beta-1} \sum_{i=1}^{n} \left[\left(\mu_A^\alpha(x_i)\mu_B^{1-\alpha}(x_i) + (1-\mu_A(x_i))^\alpha (1-\mu_B(x_i))^{1-\alpha}\right)^{\frac{\beta-1}{\alpha-1}} - 1 \right],$$
$$\alpha \neq 1,\ \alpha > 0,\ \beta \neq 1,\ \beta > 0. \tag{1.15}$$

Thereafter, Hooda and Jain [38] presented a generalized fuzzy divergence measure, its ambiguity and information improvement.

In this book, we present the generalized measures of fuzzy information and divergence. Two new parametric generalized R—norm fuzzy measures of information and two generalized fuzzy R—norm measures of divergence are presented in Chap. 2. Chapter 3 introduces the new parametric generalized exponential fuzzy divergence measure. In Chap. 4 a sequence of fuzzy mean difference divergence measures is introduced with a number of inequalities among them. One of the new generalized fuzzy divergence measure is presented in Chap. 5. A generalized Hellinger's fuzzy divergence measure is studied in Chap. 6.

1.4 Intuitionistic Fuzzy Set Theory

The notion of Atanassov's intuitionistic fuzzy sets (IFSs) was originated by Atanassov [3] which found to be well suited for dealing with both fuzziness and lack of knowledge or non-specificity. It is noted that the concept of an IFS is the best alternative approach to define a FS in cases where existing information is not enough for the definition of imprecise concepts by means of a conventional FS. Thus, the concept of Atanassov IFSs is the generalization of the concept of FSs.

In fuzzy sets, the membership value of x in X (universe of discourse) is $\mu(x)$, a single real number in [0, 1] and $1 - \mu(x)$ is taken as the non-membership value of x. But Atanassov's intuitionistic fuzzy sets have the membership value $\mu(x)$ as well as the non-membership value $v(x)$ taken into account for describing any x in X such that the sum of membership and non-membership is less than or equal to 1. Thus, an

Atanassov's intuitionistic fuzzy set is expressed with the ordered pair of real numbers $(\mu(x), v(x))$ and $(1-\mu(x)-v(x))$ is called the degree of hesitancy. Moreover, Gau and Buehrer [31] introduced the notion of vague sets. But, Bustince and Burillo [14] presented that the notion of vague sets was equivalent to that of Atanassov IFSs. The investigation on the IFS theory has been extended exponentially in last decades and applied in many areas as decision-making, medical diagnosis, pattern recognition, linguistic variables, etc.

1.4.1 Intuitionistic Fuzzy Set

An intuitionistic fuzzy set A defined on a universe of discourse $X = \{x_1, x_2, \ldots, x_n\}$ is defined as

$$A = \{\langle x_i, \mu_A(x_i), v_A(x_i)\rangle / x_i \in X\} \tag{1.16}$$

where $\mu_A : X \to [0,1]$, $v_A : X \to [0,1]$ with the condition $0 \le \mu_A + v_A \le 1 \; \forall x_i \in X$.

The numbers $\mu_A(x_i), v_A(x_i) \in [0,1]$ denote the degree of membership and non-membership of x_i to A, respectively.

For each intuitionistic fuzzy set in X we will call $\pi_A(x_i) = 1 - \mu_A(x_i) - v_A(x_i)$, the intuitionistic index or degree of hesitation of x_i in A. It is obvious that $0 \le \pi_A(x_i) \le 1$ for each $x_i \in X$. For a fuzzy set A' in X, $\pi_A(x_i) = 0$ when $v_A(x_i) = 1 - \mu_A(x_i)$. Thus, FSs are the special cases of IFSs.

1.4.2 Operations on Intuitionistic Fuzzy Sets

Atanassov [4] further defined set operations on intuitionistic fuzzy sets as follows:

Let $A, B \in \text{IFS}(X)$ given by

$$A = \{\langle x_i, \mu_A(x_i), v_A(x_i)\rangle / x_i \in X\},$$
$$B = \{\langle x_i, \mu_B(x_i), v_B(x_i)\rangle / x_i \in X\}$$

(i) $A \subseteq B$ iff $\mu_A(x_i) \le \mu_B(x_i)$ and $v_A(x_i) \ge v_B(x_i) \; \forall x_i \in X$.
(ii) $A = B$ iff $A \subseteq B$ and $B \subseteq A$.
(iii) $A^c = \{\langle x_i, v_A(x_i), \mu_A(x_i)\rangle / x_i \in X\}$.
(iv) $A \cup B = \{\langle x_i, \max(\mu_A(x_i), \mu_B(x_i)), \min(v_A(x_i), v_B(x_i))\rangle / x_i \in X\}$.
(v) $A \cap B = \{\langle x_i, \min(\mu_A(x_i), \mu_B(x_i)), \max(v_A(x_i), v_B(x_i))\rangle / x_i \in X\}$.

1.5 Intuitionistic Fuzzy Divergence Measures

As a very significant content in fuzzy mathematics, the study on the divergence measure between IFSs has received more attention in recent years. Recently, some definitions of divergence measures for IFSs have been proposed by the researchers. For example, Szmidt and Kacprzyk [80] provided the methods of distance between IFSs.

Let A and B be two intuitionistic fuzzy sets in the universe of discourse $X = \{x_1, x_2, \ldots, x_n\}$. Li [54] introduced the dissimilarity measure for intuitionistic fuzzy sets given by

$$d(A,B) = \sum_{x \in X} [|\mu_A(x) - \mu_B(x)| + |\nu_A(x) + \nu_B(x)|]/2. \tag{1.16}$$

Hung and Yang [40] defined an axiomatic structure of the divergence measure between IFSs using the Hausdorff distance as follows:

$$d_H(A,B) = \frac{1}{n} H(I_A(x_i), I_B(x_i)), \tag{1.17}$$

where $I_A(x_i)$ and $I_B(x_i)$ be subintervals on [0, 1] denoted by $I_A(x_i) = [\mu_A(x_i), 1 - \nu_A(x_i)]$ and $I_B(x_i) = [\mu_B(x_i), 1 - \nu_B(x_i)]$ and the Hausdorff distance $H(A,B) = \max\{|a_1 - b_1|, |a_2 - b_2|\}$ is defined for two intervals $A = [a_1, a_2]$ and $B = [b_1, b_2]$. and satisfies the following properties:

(i) $0 \le d_H(A,B) \le 1$
(ii) $d_H(A,B) = 0$ if and only if $A = B$.
(iii) $d_H(A,B) = d_H(B,A)$
(iv) If $A \subseteq B \subseteq C$, $A, B, C \in \text{IFSs}(X)$

Then $d_H(A,B) \le d_H(A,C)$ and $d_H(B,C) \le d_H(A,C)$.

An axiomatic definition of distance measure in IFS is also introduced by Wang and Xin [89] and applied to pattern recognition.

Vlachos and Sergiadis [88] provided an intuitionistic fuzzy divergence measure in analogy with Shang and Jiang [73] measure given by

$$D_{\text{IFS}}(A,B) = I_{\text{IFS}}(A,B) + I_{\text{IFS}}(B,A), \tag{1.18}$$

where

$$I_{\text{IFS}}(A,B) = \sum_{i=1}^{n} \left[\mu_A(x_i) \text{In} \frac{\mu_A(x_i)}{\frac{1}{2}(\mu_A(x_i) + \mu_B(x_i))} + \nu_A(x_i) \text{In} \frac{\nu_A(x_i)}{\frac{1}{2}(\nu_A(x_i) + \nu_B(x_i))} \right],$$

where $I^{\mu}(A,B) = \text{In} \frac{\mu_A(x_i)}{\mu_B(x_i)}$ is the amount of discrimination of $\mu_A(x_i)$ from $\mu_B(x_i)$.

Zhang and Jiang [100] presented a measure of divergence between IFSs/vague sets A and B as

$$D^*(A,B) = D(A,B) + D(B,A), \tag{1.19}$$

Where

$$D(A,B) = \sum_{i=1}^{n} \left[\frac{\mu_A(x_i)+1-\nu_A(x_i)}{2} \log_2 \frac{[\mu_A(x_i)+1-\nu_A(x_i)]/2}{\frac{1}{4}\{[\mu_A(x_i)+1-\nu_A(x_i)]+[\mu_B(x_i)+1-\nu_B(x_i)]\}} + \frac{1-\mu_A(x_i)+\nu_A(x_i)}{2} \log_2 \frac{[1-\mu_A(x_i)+\nu_A(x_i)]/2}{\frac{1}{4}\{[1-\mu_A(x_i)+\nu_A(x_i)]+[1-\mu_B(x_i)+\nu_B(x_i)]\}} \right]$$

Li [54] also introduced the intuitionistic fuzzy dissimilarity measure between IFSs. Hung and Yang [41] constructed J-divergence between IFSs and applied it to pattern recognition. Further, Papakostas et al. [67] provided a comparative analysis of distance and similarity measures between IFSs from a pattern recognition point of view.

Divergence measures of IFSs have been widely applied to many fields such as pattern recognition [33, 41, 67, 86], linguistic variables [40], medical diagnosis [22, 100], logical reasoning [43] and decision-making [55, 86, 87, 101]. Since the divergence measures of IFSs have been applied to many real-world situations, it is expected to have an efficient divergence measure which deals with the aspect of uncertainty, i.e. fuzziness and non specificity or lack of knowledge.

In Chap. 7 we introduce an exponential methodology for measuring the degree of divergence between two intuitionistic fuzzy sets. For this an intuitionistic fuzzy exponential divergence measure is proposed and its important properties are discussed axiomatically. In addition, the applicability and efficiency of the proposed intuitionistic fuzzy exponential divergence measure have been demonstrated by comparing it with existing intuitionistic fuzzy divergence measures using a numerical example in the framework of pattern recognition.

1.6 Decision-Making

As human beings daily face situations in which they should choose among the alternatives, thus decision-making is inherent to mankind. Decision problems have been classified in decision theory attending to their framework and elements alternatives, or ranking alternatives, from all feasible alternatives (options) based on a set of attributes or criteria [25]. It can be seen as a process composed of different phases such as information gathering, analysis and selection based on different rational and reasoning processes that lead to choose a suitable alternative among a set of possible alternatives [19, 26, 58, 60]. Thus, decision-making deals with the problem of choosing the best alternative, that is, the one with the highest degree of satisfaction for all the appropriate criteria or goals.

According to Xu and Yang [90] Multiple Criteria Decision-Making (MCDM) is an emerging discipline which supports decision-makers who face multiple and usually conflicting criteria. Moreover, it has comparatively extended up in past three decades and its extension is closely related to the development of computer science and information technology achievement particularly in compound MCDM problems.

Decision-making is the study of identifying and choosing alternatives based on the values and preferences of the decision-maker. Making a decision implies that there are alternative choices to be considered, and in such a case we want not only to identify as many of these alternatives as possible but to choose the one that: (1) has the highest probability of success or effectiveness, and (2) best fits with our goals, desires, lifestyle, values and so on. Entropy is one of the important concepts in information theory. Shannon's entropy method is appropriate for finding the suitable weight for each criterion in MADM problems [2].

MCDM is a well-established branch of decision-making that permits decision-makers to rank and select alternatives according to different criteria. Zimmermann [102] and Pirdashti et al. [69] divided MCDM into two categories: Multi-Objective Decision-Making (MODM) and Multi-Attribute Decision-Making (MADM). MODM is the same as the classical optimization models with this difference that, it is focused on optimizing of several goal functions, instead of optimizing a single goal function. However, MODM is a mathematical programming problem with multiple objective functions, whereas MADM, which is used in this study, provides several alternatives (options) according to some criteria: ranked and selected. Ranking and selecting will be made among decision alternatives described by some criteria (factors) through decision-maker knowledge and experience [24]. The MADM approach requires that the selection be made among decision alternatives described by their attributes.

Multi-Criteria Decision-Making (MCDM) has emerged as one of the fastest growing areas of operational research. A large number of MCDM methods have been developed during past three decades. Of which, we here state some of the most well known, such as: Simple Additive Weighting (SAW) method [59], Compromise Programming [93, 99], Analytic Hierarchy Process (AHP) method [72], Technique for Ordering Preference by Similarity to Ideal Solution (TOPSIS) method [42], Preference Ranking Organisation Method for Enrichment Evaluations (PROMETHEE) method [11], Grey Relational Analysis (GRA), proposed by Deng [23] as part of Grey system theory, ELimination and Choice Expressing REality (ELECTRE) method [71], Complex Proportional Assessment (COPRAS) method [98], VIKOR (VIsekriterijumska optimizacija i KOmpromisno Resenje in Serbian, means Multi-criteria Optimization and Compromise Solution) method [65], Multi-Objective Optimization on the basis of Ratio Analysis (MOORA) method [12] and Multi-Objective Optimization by Ratio Analysis plus Full Multiplicative Form (MULTIMOORA) method [13], and Additive Ratio Assessment (ARAS) method [98]. Further, a brief comparative analysis of some prominent MCDM methods is given in Stanujkic et al. [79].

A MADM problem can easily be expressed in matrix format [102, 85]. A decision matrix A is a $(M \times N)$ matrix in which element a_{ij} indicates the

Table 1.2 A typical decision matrix

Criteria	C_1	C_2	C_3	...	C_N
Alternatives	W_1	W_2	W_3	...	W_N
A_1	a_{11}	a_{12}	a_{13}	...	a_{1N}
A_2	a_{21}	a_{22}	a_{23}	...	a_{2N}
A_3	a_{31}	a_{32}	a_{33}	...	a_{3N}
$\vdots$	$\vdots$	$\vdots$	$\vdots$	$\vdots$	$\vdots$
A_M	a_{M1}	a_{M2}	a_{M3}	...	a_{MN}

performance of alternative A_i when it is evaluated in terms of decision criterion C_j, (for $i = 1, 2, 3, \ldots, M$ and $j = 1, 2, 3, \ldots, N$). It is also assumed that the decision-maker has determined the weights of relative performance of the decision criteria (denoted as W_j, for $j = 1, 2, 3, \ldots, N$). This information is best summarized in Table 1.2. Then the general MADM problem can be defined as follows.

Definition Let $A = \{A_i, i = 1, 2, 3, \ldots, M\}$ be a (finite) set of decision alternatives and $G = \{g_i, j = 1, 2, 3, \ldots, N\}$ a (finite) set of goals according to which the desirability of an action is judged. The problem is to determine the optimal alternative A^* with the highest degree of desirability with respect to all relevant goals g_i.

In the routine life sometimes we deal with unavailability of complete information, precise references and reliable data. In such situations, fuzzification of attributes plays a very important role. Thus, decision-making often occurs in a fuzzy environment where the information available is imprecise/uncertain. Bellman and Zadeh [6] developed the theory of decision behaviour in a fuzzy and intuitionistic fuzzy environment. In the literature, various methods for handling multi-criteria decision-making problems in fuzzy environment have been developed (e.g. [1, 7, 15, 17, 18, 30, 39, 44, 45, 50, 52, 53, 57, 61, 81, 83, 91, 92]).

1.6.1 *Importance of Fuzzy Set Theory in Research*

There are the some important reasons why the fuzzy set theory is applicable in different areas of research:

- Uncertainty and vagueness of such problems
- Mostly, the information that is required for development of a new model to solve the problem is not complete and measureable
- The qualitative and quantitative information is reduced due to the inexact information and biased estimation of the people concerned.

In view of importance of fuzzy set theory we develop the different methods to solve decision-making problems in fuzzy environment utilizing the proposed generalized fuzzy divergence measures. Although a lot of multi-criteria decision-making (MCDM) methods are now available to deal with varying evaluation and selection problems, in the present study we explore the applicability of new MCDM methods that are simple, efficient and computationally easy. For this, a method of strategic decision-making is developed using the proposed new parametric exponential fuzzy divergence measure and a comparative study of the proposed method of strategic decision-making with the existing methods in fuzzy environment is also presented with discussion in Chap. 3. Chapter 5 proposes an efficient and simple method for solving decision-making problems in fuzzy environment using the proposed new generalized fuzzy divergence measure. The proposed generalized Hellinger's fuzzy divergence measure is utilized in Chap. 6 to develop a new method for solving decision-making problems with ease to computation. Further, a method to solve multi-attribute decision-making problem is developed using the proposed exponential divergence measure in intuitionistic fuzzy environment with an illustrative example. A comparative study of the proposed method with the existing TOPSIS and MOORA methods of multi-attribute decision-making in an intuitionistic fuzzy environment is presented in Chap. 7.

1.7 Pattern Recognition

Pattern recognition is one of the problems that need to recognize that the given pattern belongs to what class, among several available classes. The process of the pattern recognition in fuzzy environment by use of an algorithm is given below.

We here present an algorithm to recognize and classify the given pattern in the pattern recognition problems given in Vlachos and Sergiadis [88].

Suppose that we are given m known patterns $P_1, P_2, P_3, \ldots, P_m$ which have classifications $C_1, C_2, C_3, \ldots, C_m$, respectively. The patterns are represented by the following fuzzy sets in the universe of discourse $X = \{x_1, x_2, \ldots, x_n\}$:

$$P_i = \{\langle x_j, \mu_{P_i}(x_j)\rangle / x_j \in X\},$$

where $i = 1, 2, \ldots, m$ and $j = 1, 2, \ldots, n$.

Given an unknown pattern Q, represented by the fuzzy set

$$Q_i = \{\langle x_j, \mu_{Q_i}(x_j)\rangle / x_j \in X\}.$$

Our aim here is to classify Q to one of the classes $C_1, C_2, C_3, \ldots, C_m$. According to the principle of minimum divergence/discrimination information between fuzzy sets, the process of assigning Q to C_{k^*} is described by

$$k^* = \arg\min_k \{D(P_k, Q)\}.$$

According to this algorithm, the given pattern can be recognized so that the best class can be selected. It is a practical application of minimum divergence measure principle of Shore and Gray [76] to pattern recognition.

This algorithm is used by us in Chap. 4 to show that Q has been classified to same class correctly by using all of the proposed fuzzy mean difference divergence measures. This algorithm is also applied in Chaps. 5 and 6 to illustrate that the proposed generalized fuzzy divergence measures are applicable in pattern recognition and medical diagnosis problems. Moreover, the above algorithm is used in Chap. 7 to demonstrate the efficiency and applicability of the proposed intuitionistic fuzzy exponential divergence measure by comparing it with the existing intuitionistic fuzzy divergence measures.

1.8 Linguistic Variables

There are situations in which the information can be measured accurately in a qualitative form but may not be in a quantitative one, and thus, the use of a linguistic approach becomes essential. For example, we are often lead to use words in natural language instead of numerical values in the phenomena related to human perception. To represent the qualitative aspect as linguistic values by means of linguistic variables is an approximate technique of linguistic approach. Thus, the linguistic variables play a key role in linguistic approach.

Zadeh [95] introduced the concept of linguistic variables and provided a direct way to represent the linguistic information by means of linguistic variables. Variables whose values are not numbers, but words or sentences in natural or artificial languages are called linguistic variables [35, 36]. For example, a numerical variable takes numerical value: age = 25 years, a linguistic variable takes the linguistic value: age is young, thus, a linguistic value is a fuzzy set.

Chen and Hwang [16] pointed out that there are situations in which the information may be unquantifiable due to its nature, and thus, it may be stated only in linguistic terms. Herrera and Herrera [36] provided an example, '(when evaluating the 'comfort' or 'design' of a car terms like 'good', 'medium', 'bad' can be used). In other cases, precise quantitative information may not be stated because either it is unavailable or the cost of its computation is too high, so an 'approximate value' may be tolerated. For example, when evaluating the speed of a car, linguistic terms like 'fast', 'very fast', 'slow' may be used instead of numerical values.

We deal with the construction of generalized fuzzy information measures and their application to different areas in our environment. Thus, the efficiency of these generalized fuzzy information measures in linguistic variables is also required. For this, in Chap. 4 we develop a sequence of fuzzy mean difference divergence measures and present their efficiency to linguistic variables.

1.9 Concluding Remarks

In this chapter, we have laid down the foundations for this book's motivation, and have given a high-level overview of its main aim: to present the applications of the fresh generalizations of fuzzy information and divergence measures. First, the extant literature offers several fuzzy information and divergence measures. But, there is still much scope for development of the better information-theoretic measures which will deal with the real-world problems in pattern recognition, image processing, speech recognition, bioinformatics, fuzzy aircraft control, feature selection, etc. Second, the generalization of information measures improves their flexibility from the application point of view. Third, generalized measures have capability to properly and efficiently handle the information which deals with the different aspects of uncertainty.

The rest of this book is devoted to fleshing out the generalizations of fuzzy information measures proposed, and to illustrate these with a number of illustrative examples. Chapter 2 provides two parametric generalizations of *R*—norm information and divergence measures and establishes a relation between them. A generalized exponential measure of fuzzy divergence and a method to solve the problem related to strategic decision-making are given in Chap. 3. Thereafter, a sequence of fuzzy divergence measures with their applicability to deal with linguistic variables and dynamic pattern recognition is presented with discussion in Chap. 4. In what follows, Chap. 5 relates a newly generalized fuzzy divergence measure and method to solve the multi-criteria decision-making problems with the help of illustrative examples.

In this way, generalization of Hellinger's fuzzy divergence measure with its application to multi-criteria decision-making and medical diagnosis is studied in Chap. 6. An efficient and novel generalized exponential intuitionistic fuzzy divergence measure and a method to solve multi-attribute decision-making are introduced in Chap. 7. Finally, this book concludes by discussing outlook of other application domains of proposed fuzzy generalized measures and some possible further developments of this line of research.

References

1. Amiri-Aref M, Javadian N, Kazemi M (2012) A new fuzzy positive and negative ideal solution for fuzzy TOPSIS. Wseas Trans Circ Syst 11(3):92–103
2. Andreica ME, Dobre I, Andreica MI, Resteanu C (2010) A new portfolio selection method based on interval data. Stud Inform Control 19(3):253–262
3. Atanassov KT (1986) Intuitionistic fuzzy sets. Fuzzy Sets Syst 20:87–96
4. Atanassov KT (1994) New operations defined over the intuitionistic fuzzy sets. Fuzzy Sets Syst 61:137–142
5. Bajaj RK, Hooda DS (2010) On some new generalized measures of fuzzy information. World Acad Sci Eng Technol 62:747–753
6. Bellman RE, Zadeh LA (1970) Decision-making in a fuzzy environment. Manage Sci 17:141–164
7. Beynon M, Cosker D, Marshall D (2001) An expert system for multi-criteria decision making using dempster shafer theory. Expert Syst Appl 20(4):357–367
8. Bhandari D, Pal NR (1993) Some new information measures for fuzzy sets. Inf Sci 67 (3):209–228

9. Bhatia PK, Singh S (2012) Three families of generalized fuzzy directed divergence. AMO-Adv Model Optim 14(3):599–614
10. Bouchon-Meunier B, Rifqi M, Bothorel S (1996) Towards general measures of comparison of objects. Fuzzy Sets Syst 84:143–153
11. Brans JP, Vincke P (1985) A preference ranking organization method: The PROMETHEE method for MCDM. Manage Sci 31(6):647–656
12. Brauers WKM, Zavadskas EK (2006) The MOORA method and its application to privatization in transition economy. Control Cybern 35(2):443–468
13. Brauers WKM, Zavadskas EK (2010) Project management by MULTIMOORA as an instrument for transition economies. Tech Econ Dev Econ 16(1):5–24
14. Bustince H, Burillo P (1996) Vague sets are intuitionisic fuzzy sets. Fuzzy Sets Syst 79: 403–405
15. Chen SJ, Chen SM (2005) A prioritized information fusion method for handling fuzzy decision-making problems. Appl Intel 22(3):219–232
16. Chen SJ, Hwang CL (1992) Fuzzy multiple attribute decision making-methods and applications. Springer, Berlin
17. Chen SM, Lee LW (2010) Fuzzy multiple attributes group decision-making based on the interval type-2 TOPSIS method. Expert Syst Appl 37(4):2790–2798
18. Chen SM, Wang CH (2009) A generalized model for prioritized multi-criteria decision making systems. Expert Syst Appl 36(3):4773–4783
19. Clement RT (1996) Making hard decisions. An introduction to decision analysis. Duxbury Press, Pacific Grove
20. Couso I, Janis V, Montes S (2000) Fuzzy divergence measures. Acta Univ M Belii 8:21–26
21. De Luca A, Termini S (1972) A definition of non-probabilistic entropy in the setting of fuzzy set theory. Inf Control 20(4):301–312
22. De SK, Biswas R, Roy AR (2001) An application of intuitionistic fuzzy sets in medical diagnosis. Fuzzy Sets Syst 117(2):209–213
23. Deng JL (1989) Introduction to grey system. J Grey Syst 1(1):1–24
24. Devi K, Yadav SP, Kumar SP (2009) Extension of fuzzy TOPSIS method based on vague sets. Comput Cogn 7:58–62
25. Duncan R, Raiffa H (1985) Games and decision. Introduction and critical survey. Dover Publications
26. Evangelos T (2000) Multi-criteria decision making methods: a comparative study. Kluwer Academic Publishers, Dordrecht
27. Fan J, Xie W (1999) Distance measures and induced fuzzy entropy. Fuzzy Sets Syst 104 (2):305–314
28. Félix P, Barroa S, Marín R (2003) Fuzzy constraint networks for signal pattern recognition. Artif Intell 148:103–140
29. Ferreri C (1980) Hyperentropy and related heterogeneity divergence and information measures. Statistica 40(2):155–168
30. Fu G (2008) A fuzzy optimization method for multi-criteria decision making: an application to reservoir flood control operation. Expert Syst Appl 34(1):145–149
31. Gau WL, Buehrer DJ (1993) Vague sets. IEEE Trans Syst Man Cybern 23:610–614
32. Hartley RVL (1928) Transmission of information. Bell Syst Tech J 7(3):535–563
33. Hatzimichailidis AG, Papakostas GA, Kaburlasos VG (2012) A novel distance measure of intuitionistic fuzzy sets and its application to pattern recognition problems. Int J Intel Syst 27(4):396–409
34. Havrada JH, Charvat F (1967) Quantification methods of classification processes: concept of structural α-entropy. Kybernetika 3(1):30–35
35. Herrera F, Herrera-Viedma E (1996) A model of consensus in group decision making under linguistic assessments. Fuzzy Sets Syst 78(1):73–87
36. Herrera F, Herrera-Viedma E (2000) Linguistic decision analysis: steps for solving decision problems under linguistic information. Fuzzy Sets Syst 11(1):67–82
37. Hooda DS (2004) On generalized measures of fuzzy entropy. Math Slovaca 54:315–325

38. Hooda DS, Jain D (2012) The generalized fuzzy measures of directed divergence, total ambiguity and information improvement. Investig Math Sci. 2:239–260
39. Hua Z, Gong B, Xu X (2008) A DS—AHP approach for multi-attribute decision making problem with incomplete information. Expert Syst Appl 34(3):2221–2227
40. Hung WL, Yang MS (2004) Similarity measures of intuitionistic fuzzy sets based on Hausdorff distance. Pattern Recogn Lett 25:1603–1611
41. Hung WL, Yang MS (2008) On the J-divergence of intuitionistic fuzzy sets and its application to pattern recognition. Inf Sci 178(6):1641–1650
42. Hwang CL, Yoon K (1981) Multiple attribute decision making—methods and applications. Springer, New York
43. Jiang YC, Tang Y, Wang J, Tang S (2009) Reasoning within intuitionistic fuzzy rough description logics. Inf Sci 179:2362–2378
44. Joshi D, Kumar S (2014) Intuitionistic fuzzy entropy and distance measure based TOPSIS method for multi-criteria decision making. Egypt Inf J 15:97–104
45. Kahraman C, Cebi S (2009) A new multi-attribute decision making method: hierarchical fuzzy axiomatic design. Expert Syst Appl 36(3):4848–4861
46. Kapur JN (1997) Measures of fuzzy information. Math Sci Trust Soc, New Delhi
47. Kaufmann A (1980) Fuzzy subsets: fundamental theoretical elements. Academic Press, New York 3
48. Klir GJ, Folger TA (2009) Fuzzy sets, uncertainty and information. Prentice Hall, New York
49. Kosko B (1990) Fuzziness vs. probability. Int J Gen Syst 17:211–240
50. Kulak O (2005) A decision support system for fuzzy multi-attribute selection of material handling equipments. Expert Syst Appl 29(2):310–319
51. Kullback S, Leibler RA (1951) On information and sufficiency. Ann Math Stat 22(1):79–86
52. Kuo MS, Tzeng GH, Huang WC (2007) Group decision-making based on concepts of ideal and anti-ideal points in a fuzzy environment. Math Comput Model 45:324–339
53. Kwon O, Kim M (2004) My message: case-based reasoning and multi-criteria decision making techniques for intelligent context-aware message filtering. Expert Syst Appl 27(3): 467–480
54. Li DF (2004) Some measures of dissimilarity in intuitionistic fuzzy structures. J Comput Syst Sci 68(1):115–122
55. Li DF (2005) Multi-attribute decision-making models and methods using intuitionistic fuzzy sets. J Comput Syst Sci 70(1):73–85
56. Lin J (1991) Divergence measure based on Shannon entropy. IEEE Trans Inf Theory 37(1): 145–151
57. Lin HY, Hsu PY, Sheen GJ (2007) A fuzzy-based decision-making procedure for data warehouse system selection. Expert Syst Appl 32(3):939–953
58. Lu J, Zhang G, Ruan D (2007) Multi-objective group decision making. Methods, software and applications with fuzzy set techniques. Imperial College Press
59. MacCrimon KR(1968) Decision marking among multiple-attribute alternatives: a survey and consolidated approach. RAND memorandum, RM-4823-ARPA. The Rand Corporation, S Martínez anta Monica, California
60. Martínez L (2010) Computing with words in linguistic decision making: Analysis of Linguistic Computing Models. IEEE 5–8
61. Mikhailov L (2003) Deriving priorities from fuzzy pair wise comparison judgments. Fuzzy Sets Syst 134:365–385
62. Montes S, Couso I, Gil P, Bertoluzza C (2002) Divergence measure between fuzzy sets. Int J Approximate Reasoning 30:91–105
63. Nyquist H (1924) Certain factors affecting telegraph speed. Bell Syst Tech J 3(2):324–346
64. Nyquist H (1928) Certain topics in telegraph transmission theory. A.I.E.E. Trans 47: 617–644
65. Opricovic S (1998) Visekriterijumska optimizacija u građevinarstvu—multi-criteria optimization of civil engineering systems. Faculty of Civil Engineering, Belgrade (Serbian)

66. Pal NR, Pal SK (1989) Object background segmentation using new definition of entropy. IEE Proc Comput Digital Tech 136(4):248–295
67. Papakostas GA, Hatzimichailidis AG, Kaburlasos VG (2013) Distance and similarity measures between intuitionistic fuzzy sets: a comparative analysis from a pattern recognition point of view. Pattern Recogn Lett 34(14):1609–1622
68. Parkash O, Sharma PK, Kumar S (2006) Two new measures of fuzzy divergence and their properties. SQU J Sci 11:69–77
69. Pirdashti M, Ghadi A, Mohammadi M, Shojatalab G (2009) Multi-criteria decision-making selection model with application to chemical engineering management decisions. World Academy Sci Eng Technol 3(1):27–32
70. Renyi A (1961) On measures of entropy and information. In: Proceeding of Fourth Berkeley Symposium on Mathematics, Statistics and Probability, vol 1, 547–561
71. Roy B (1991) The outranking approach and the foundation of ELECTRE methods. Theor Decis 31:49–73
72. Saaty TL (1980) The analytic hierarchy process: planning, priority setting, resource allocation. McGraw-Hill, New York
73. Shang X, Jiang G (1997) A note on fuzzy information measures. Pattern Recogn Lett 18 (5):425–432
74. Shannon CE (1948) The mathematical theory of communication. Bell Syst Tech J 27(3): 379–423
75. Sharma BD, Mittal DP (1977) New non-additive measures of relative information. J Comb Inf Syst Sci 2:122–133
76. Shore JE, Gray RM (1982) Minimization cross-entropy pattern classification and cluster analysis. IEEE Trans Pattern Anal Mach Intel 4(1):11–17
77. Singh RP, Tomar VP (2010) Refinement of inequalities among fuzzy divergence measures. Adv Appl Res 2(2):142–156
78. Singpurwalla ND, Booker JM (2004) Membership functions and probability measures of fuzzy sets. J Am Stat Assoc 99:867–877
79. Stanujkic D, Đorđević B, Đorđević M (2013) Comparative analysis of some prominent mcdm methods: a case of ranking serbian banks. Serbian J Manag 8(2):211–241
80. Szmidt E, Kacprzyk J (2000) Distances between intuitionistic fuzzy sets. Fuzzy Sets Syst 114(3):505–518
81. Tacker EC, Silvia MT (1991) Decision making in complex environments under conditions of high cognitive loading: a personal expert systems approach. Expert Syst Appl 2(2–3): 121–127
82. Taneja IJ (2008) On mean divergence measures. In: Barnett NS, Dragomir SS (eds) Advances in Inequalities from probability theory and statistics. Nova, USA, pp 169–186
83. Tomar VP, Ohlan A (2014a) New parametric generalized exponential fuzzy divergence measure. J Uncertain Anal Appl 2(1):1–14
84. Tomar VP, Ohlan A (2014b) Sequence of fuzzy divergence measures and inequalities. Adv Model Optim 16(2):439–452
85. Triantaphyllou E, Shu B, Nieto Sanchez S, Ray T (1998) Encyclopedia of electrical and electronics engineering, Webster JG (ed), John Wiley and Sons, New York, NY, 15, 175–186
86. Verma R, Sharma BD (2011) On generalized exponential fuzzy entropy. World Acad Sci Eng Technol 69:1402–1405
87. Verma R, Sharma BD (2012) On generalized intuitionistic fuzzy divergence (relative information) and their properties. J Uncertain Syst 6(4):308–320
88. Vlachos IK, Sergiadis GD (2007) Intuitionistic fuzzy information-application to pattern recognition. Pattern Recognit Lett 28(2):197–206
89. Wang WQ, Xin XL (2005) Distance measures between intuitionistic fuzzy sets. Pattern Recogn Lett 26:2063–2069
90. Xu L, Yang JB (2001) Introduction to multi-criteria decision making and the evidential reasoning approach. University of Manchester, Institute of Science and Technology, Manchester School of Management. ISBN 1 86115 111 X

91. Yager RR (1991) Non-monotonic set theoretic operations. Fuzzy Sets Syst 42(2):173–190
92. Yager RR (1992) Second order structures in multi-criteria decision making. Int J Man Mach Stud 36(6):553–570
93. Yu PL (1973) A class of solutions for group decision problems. Manage Sci 19(8):936–946
94. Zadeh LA (1965) Fuzzy sets. Inf Control 8:338–353
95. Zadeh LA (1968) Probability measures of fuzzy events. J Math Anal Appl 23:421–427
96. Zadeh LA (1987) Fuzzy sets and applications: Selected Papers by L.A. Zadeh. Wiley, New York
97. Zadeh LA (1995) Discussion: probability theory and fuzzy logic are complementary rather than competitive. Technometrics 37(3):271–276
98. Zavadskas EK, Kaklauskas A, Sarka V (1994) The new method of multicriteria complex proportional assessment of projects. Technol Econ Dev Econ 1(3):131–139
99. Zeleny M (1973) Compromise programming. In: Cochrane JL, Zeleny M (eds) Multiple criteria decision making. University of South Carolina Press, Columbia, SC, 262–301
100. Zhang QS, Jiang SY (2008) A note on information entropy measures for vague sets and its applications. Inf Sci 178:4184–4191
101. Zhang SF, Liu SY (2011) A GRA-based intuitionistic multi-criteria decision making method for personnel selection. Expert Syst Appl 38(9):11401–11405
102. Zimmermann HJ (1991) Fuzzy set theory and its applications. Kluwer Academic Publishers, Second Edition, Boston, MA

Chapter 2
Parametric Generalized *R*-norm Fuzzy Information and Divergence Measures

This chapter deals with two new generalized *R*-norm fuzzy entropy measures and some of their interesting properties. In addition, we propose two new generalized *R*-norm fuzzy divergence measures. In order to check the validity of proposed measures, their essential properties are studied.

Entropy is very important for measuring uncertain information first introduced by Shannon [14] to measure the uncertain degree of the randomness in a probability distribution. Let X is a discrete random variable with probability distribution $P = (p_1, p_2, \ldots, P_n)$ in an experiment. The information contained in this experiment is given by

$$H(P) = -\sum_{i=1}^{n} p_i \log p_i \tag{2.1}$$

which is the well-known Shannon [14] entropy.

Fuzziness, a feature of uncertainty, is found in our verbal communication, in our judgment and in the way we process information. The fuzzy entropy is one of important digital features of fuzzy sets and occupies an important place in system model and system design. It is an important concept for measuring fuzzy information. Fuzziness, an element of uncertainty, results from the lack of sharp difference of being or not being a member of the set. A measure of fuzziness used and citied in literature is called the fuzzy entropy, first mentioned by Zadeh [15]. It has wide applications in the area of image processing, pattern recognition, speech recognition, decision-making, etc.

In Sect. 2.1 we review the definition of fuzzy information measure, *R*-norm fuzzy measures of information and there generalization existing in the literature. Two new parametric generalizations of *R*-norm measure of fuzzy information are defined and the essential properties are proved to check their authenticity in Sect. 2.2. In order to make a comparison between the proposed generalized *R*-norm measures of fuzzy information with one existing *R*-norm fuzzy information

A. Ohlan and R. Ohlan, *Generalizations of Fuzzy Information Measures*,
DOI 10.1007/978-3-319-45928-8_2

measure, a numerical example is given in Sect. 2.3. In Sect. 2.4 we review some existing R-norm fuzzy measures of divergence. Section 2.5 introduces and validates two new parametric generalized R-norm fuzzy divergence measures. Some of interesting properties of these two proposed fuzzy divergence measures are established and proved in Sect. 2.6. A relation between the new generalized R-norm measures of fuzzy information and divergence is established in Sect. 2.7. The concluding remarks are drawn in the last section.

2.1 Generalized *R*-norm Fuzzy Information Measures

We begin by reviewing the definition of fuzzy information measure, R-norm fuzzy measures of information and their generalization existing in the literature. De Luca and Termini [4] introduced the measure of fuzzy entropy corresponding to Shannon [14] entropy given in Eq. 2.1 as

$$H(A) = -\sum_{i=1}^{n} [\mu_A(x_i)\log\mu_A(x_i) + (1-\mu_A(x_i))\log(1-\mu_A(x_i))] \tag{2.2}$$

satisfying the following essential properties:

(P1) $H(A)$ is minimum if and only if A is a crisp set, i.e. $\mu_A(x_i) = 0$ or 1 for all $x_i : i = 1, 2, \ldots n$.
(P2) $H(A)$ is maximum if and only if A is most fuzzy set, i.e. $\mu_A(x_i) = 0.5$ for all $x_i : i = 1, 2, \ldots n$.
(P3) $H(A) \geq H(A^*)$, where A^* is sharpened version of A.
(P4) $H(A) = H(\overline{A})$, where $\overline{A}$ is the complement of A.

Later on Bhandari and Pal [2] defined the following exponential fuzzy entropy corresponding to Pal and Pal [13] exponential entropy as

$$E(A) = \frac{1}{n(\sqrt{e}-1)}\sum_{i=1}^{n}\left[\mu_A(x_i)e^{(1-\mu_A(x_i))} + (1-\mu_A(x_i))e^{\mu_A(x_i)} - 1\right] \tag{2.3}$$

Boekee and Lubbe [3] defined and studied R-norm information measure, which has been mentioned by Arimoto [1] as an example of a generalized class of information measures of the distribution P for $R \in \mathbf{R}$ as given by

$$H_R(P) = \frac{R}{R-1}\left[1 - (\sum_{i=1}^{n} p_i^R)^{\frac{1}{R}}\right]; \quad R > 0, R \neq 1. \tag{2.4}$$

The R-norm information measure (2.4) is a real function $\Delta_n \to \mathbf{R}$, defined on Δ_n (Set of all nary probability distributions which satisfies the condition $p_i \geq 0, i = 1,2,3,\ldots,n$ and $\sum_{i=1}^{n} p_i = 1$) where $n \geq 2$ and $\mathbf{R}$ is the set of real numbers.

The most interesting property of this measure is that when $R \to 1$, R-norm information measure (2.4) approaches to Shannon's entropy (2.1) and in case $R \to \infty$, $H_R(P) \to (1 - \max p_i)$, $i = 1,2,\ldots,n$.

Analogous to measure (2.4) Hooda [5] proposed the following R-norm fuzzy measure of information

$$H_R(A) = \frac{R}{R-1}\sum_{i=1}^{n}\left[1 - (\mu_A^R(x_i) + (1-\mu_A(x_i))^R)^{\frac{1}{R}}\right];\ R(>0) \neq 1. \qquad (2.5)$$

Further, from the significant studies it is noted that Hooda and Bajaj [6] and Hooda and Jain [7] provide the generalization of R-norm fuzzy information measure (2.5) corresponding to generalized R-norm information measures proposed by Hooda and Ram [8] and Hooda and Sharma [9] respectively.

Kumar [11] generalized the measure (2.4) and gave R-norm measure of information of order α which is

$$H_R^{\alpha}(P) = \frac{R}{R-\alpha}\left[1 - \left(\sum_{i=1}^{n} p_i^{\frac{R}{\alpha}}\right)^{\frac{\alpha}{R}}\right]; \quad 0<\alpha\leq 1,\ R(>0) \neq 1. \qquad (2.6)$$

In addition, Kumar and Choudhary [12] generalized the measure (2.4) and gave the R-norm information measure of degree m as

$$H_R^m(P) = \frac{R-m+1}{R-m}\left[1 - \left(\sum_{i=1}^{n} p_i^{R-m+1}\right)^{\frac{1}{R-m+1}}\right]; \qquad (2.7)$$
$$R-m+1>0,\ R\neq m,\ R,m>0(\neq 1).$$

2.2 New Parametric Generalized *R*-norm Fuzzy Information Measures

Let A be the fuzzy set defined in a universe of discourse $X = \{x_1, x_2, \ldots, x_n\}$ having the membership values $\mu_A(x_i)$, $i = 1,2,\ldots,n$. Corresponding to measures given by Kumar [11] (2.6) and Kumar and Choudhary [12] (2.7), the following generalized R-norm fuzzy information measures are proposed as follows:

$$H_R^\alpha(A) = \frac{R}{R-\alpha}\sum_{i=1}^{n}\left[1 - \left\{\mu_A^{\frac{R}{\alpha}}(x_i) + (1-\mu_A(x_i))^{\frac{R}{\alpha}}\right\}^{\frac{\alpha}{R}}\right];\quad 0<\alpha\leq 1,\ R(>0)\neq 1. \tag{2.8}$$

$$H_R^m(A) = \frac{R-m+1}{R-m}\sum_{i=1}^{n}\left[1 - \left\{\mu_A^{R-m+1}(x_i) + (1-\mu_A(x_i))^{R-m+1}\right\}^{\frac{1}{R-m+1}}\right]; \tag{2.9}$$

$$R-m+1>0,\ R\neq m,\ R,m>0(\neq 1).$$

Theorem 2.1 *The generalized R-norm fuzzy information measures* $H_R^\alpha(A)$ *and* $H_R^m(A)$ *are valid measures of fuzzy information.*

Proof **P1 (Sharpness)**: The measures (2.8) and (2.9) clearly satisfy the property P1, i.e. $H_R^\alpha(A) = 0$ and $H_R^m(A) = 0$ if and only if A is non-fuzzy set or crisp set.

P2 (Maximality): To verify that the proposed measure (2.8) is concave; the values of $H_R^\alpha(A)$ are computed first for a fixed value of R and different values of α, second for a fixed value of α and different values of R.

Case 1: Let us assume a particular value of $R = 0.5$ and different values of α. The computed values of $H_R^\alpha(A)$ using (2.8) with different values of α are given in Table 2.1 which show the concavity of $H_R^\alpha(A)$ with respect to α.

A glance at Fig. 2.1 more precisely shows the concave behaviour of $H_R^\alpha(A)$ with respect to α.

Case 2: Let us assume a particular value of $\alpha = 0.7$ and different values of R. The computed values of $H_R^\alpha(A)$ for $\alpha = 0.7$ using (2.8) and different values of R are given in Table 2.2.

Table 2.2 and Fig. 2.2 clearly depict the concavity of $H_R^\alpha(A)$ with respect to R using (2.8).

Thus $H_R^\alpha(A)$ is a concave function with respect to α and R and its maximum value exists at $\mu_A(x_i) = 0.5$.

Now to verify that the proposed measure $H_R^m(A)$ in (2.9) is concave; the values of $H_R^m(A)$ first for a fixed value of R and different values of m, second for a fixed value of m and different values of R are computed.

Case 1: Let us assume a particular value of $R = 0.5$ and different values of m. The computed values of $H_R^m(A)$ using (2.9) with different values of m are given in Table 2.3 and Fig. 2.3 which depict the concavity of $H_R^m(A)$ using (2.9) with respect to m.

Case 2: Let us assume a particular value of $m = 0.7$ and different values of R. The computed values of $H_R^m(A)$ for $m = 0.7$ using (2.9) and different values of R are given in Table 2.4.

It may be seen from the results presented in Table 2.4 and Fig. 2.4 that behaviour of $H_R^m(A)$ with respect to R is concave.

Table 2.1 Different values of $H_R^{\alpha}(A)$ for $R = 0.5$ and for different values of α

α	$\mu_A(x_i)$	0	0.1	0.2	0.3	0.4	0.5	0.6	0.7	0.8	0.9	1
0.1	$H_{0.5}^{0.1}(A)$	0	0.1250	0.2498	0.3725	0.4812	0.7179	0.4812	0.3725	0.2498	0.1250	0
0.3	$H_{0.5}^{0.3}(A)$	0	0.2155	0.3832	0.5054	0.5802	0.6053	0.5802	0.5054	0.3832	0.2155	0
0.4	$H_{0.5}^{0.4}(A)$	0	0.2705	0.4437	0.5588	0.6254	0.6472	0.6254	0.5588	0.4437	0.2705	0
0.7	$H_{0.5}^{0.7}(A)$	0	0.4319	0.6124	0.7204	0.7797	0.7988	0.7797	0.7204	0.6124	0.4319	0
1.0	$H_{0.5}^{1.0}(A)$	0	0.6	0.8	0.9165	0.9798	1	0.9798	0.9165	0.8	0.6	0

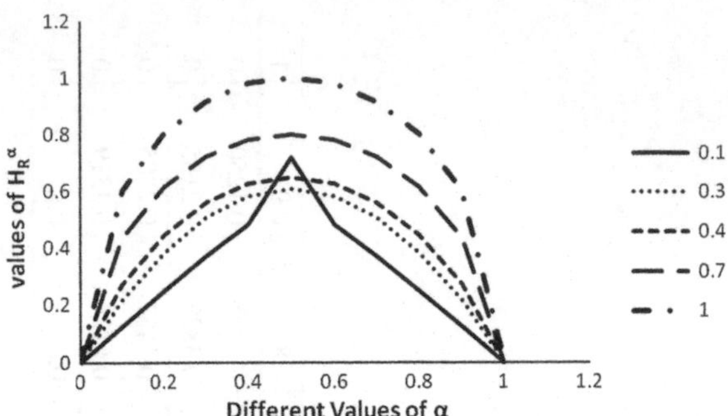

Fig. 2.1 Concave behaviour of $H_R^\alpha(A)$ with respect to α

Thus $H_R^m(A)$ is a concave function with respect to m and R and its maximum value exists at $\mu_A(x_i) = 0.5$.

Hence the measures (2.8) and (2.9) satisfy the property P2.

P3 (Resolution): Since $H_R^\alpha(A)$ and $H_R^m(A)$ are increasing functions of $\mu_A(x_i)$ in the range $[0, 0.5)$ and decreasing function in the range $(0.5, 1]$, therefore

$$\begin{aligned} &\mu_{A^*}(x_i) \le \mu_A(x_i) \Rightarrow H_R^\alpha(A^*) \le H_R^\alpha(A) \text{ in } [0, 0.5) \\ \text{and} \quad &\mu_{A^*}(x_i) \ge \mu_A(x_i) \Rightarrow H_R^\alpha(A^*) \ge H_R^\alpha(A) \text{ in } (0.5, 1] \end{aligned}$$

Taking the above equations together, it comes

$$H_R^\alpha(A^*) \le H_R^\alpha(A).$$

Similarly, $H_R^m(A^*) \le H_R^m(A)$.

P4 (Symmetry): From the definition of $H_R^\alpha(A)$ and $H_R^m(A)$ and with $\mu_{\overline{A}}(x_i) = 1 - \mu_A(x_i)$, it is obvious that $H_R^\alpha(\overline{A}) = H_R^\alpha(A)$ and $H_R^m(\overline{A}) = H_R^m(A)$.

Hence in view of the definition of fuzzy information measure of De Luca and Temini [4] provided in Sect. 2.1, $H_R^\alpha(A)$ and $H_R^m(A)$ are valid measures of fuzzy information. The measure (2.8) can be called as the generalized $\alpha - R$ -norm fuzzy information measure and measure (2.9) as the generalized R-norm fuzzy information measure of type R and degree m.

Limiting and Particular Cases:

(i) When $\alpha = 1$ and $m = 1$, (2.8) and (2.9) reduce to $H_R(A)$ in (2.5).
(ii) When $\alpha = 1$, $m = 1$ and $R \to 1$, (2.8) and (2.9) reduce to $H(A)$ in (2.2).
(iii) When $\alpha = 1$, $m = 1$ and $R \to \infty$, (2.8) and (2.9) reduce to

$$\sum_{i=1}^{n} [1 - \max\{\mu_A(x_i), 1 - \mu_A(x_i)\}].$$

Table 2.2 Different values of $H_R^{\alpha}(A)$ for $\alpha = 0.7$ and for different values of R

R	$\mu_A(x_i)$	0	0.1	0.2	0.3	0.4	0.5	0.6	0.7	0.8	0.9	1
0.1	$H_{0.1}^{0.7}(A)$	0	6.8071	8.6641	9.7356	10.3152	10.5	10.3152	9.7356	8.6641	6.8071	0
0.2	$H_{0.2}^{0.7}(A)$	0	1.2087	1.5379	1.7276	1.8301	1.8627	1.8301	1.7276	1.5379	1.2087	0
0.4	$H_{0.4}^{0.7}(A)$	0	0.5275	0.7174	0.8289	0.8896	0.9090	0.8896	0.8289	0.7174	0.5275	0
0.6	$H_{0.6}^{0.7}(A)$	0	0.3698	0.5467	0.6552	0.7153	0.7348	0.7153	0.6552	0.5467	0.3698	0
0.8	$H_{0.8}^{0.7}(A)$	0	0.2911	0.4652	0.5783	0.6429	0.6640	0.6429	0.5783	0.4652	0.2911	0

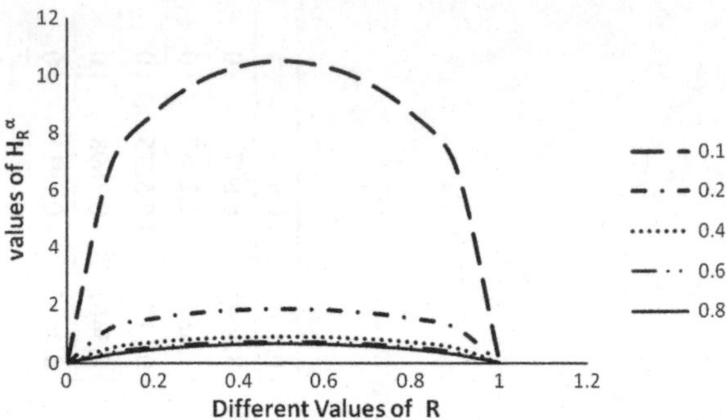

Fig. 2.2 Concave behaviour of $H_R^\alpha(A)$ with respect to R

Theorem 2.2 *For* $A, B \in FS(X)$, $H_R^\alpha(A \cup B) + H_R^\alpha(A \cap B) = H_R^\alpha(A) + H_R^\alpha(B)$.

Proof Let us consider the sets

$$X_1 = \{x/\, x \in X,\ \mu_A(x_i) \geq \mu_B(x_i)\} \tag{2.10}$$

$$X_2 = \{x/\, x \in X,\ \mu_A(x_i) < \mu_B(x_i)\} \tag{2.11}$$

where $\mu_A(x_i)$ and $\mu_B(x_i)$ be the fuzzy membership functions of A and B respectively.

$$\begin{aligned} H_R^\alpha(A \cup B) &= \frac{R}{R-\alpha}\sum_{i=1}^{n}\left[1 - \left(\mu_{A\cup B}^{\frac{R}{\alpha}}(x_i) + (1-\mu_{A\cup B}(x_i))^{\frac{R}{\alpha}}\right)^{\frac{\alpha}{R}}\right] \\ &= \frac{R}{R-\alpha}\left[\begin{array}{l} \sum_{X_1}[1 - (\mu_A^{\frac{R}{\alpha}}(x_i) + (1-\mu_A(x_i))^{\frac{R}{\alpha}})^{\frac{\alpha}{R}}] \\ \quad + \sum_{X_2}[1 - (\mu_B^{\frac{R}{\alpha}}(x_i) + (1-\mu_B(x_i))^{\frac{R}{\alpha}})^{\frac{\alpha}{R}}] \end{array}\right] \end{aligned} \tag{2.12}$$

$$\begin{aligned} H_R^\alpha(A \cap B) &= \frac{R}{R-\alpha}\sum_{i=1}^{n}\left[1 - \left(\mu_{A\cap B}^{\frac{R}{\alpha}}(x_i) + (1-\mu_{A\cap B}(x_i))^{\frac{R}{\alpha}}\right)^{\frac{\alpha}{R}}\right] \\ &= \frac{R}{R-\alpha}\left[\begin{array}{l} \sum_{X_1}[1 - (\mu_B^{\frac{R}{\alpha}}(x_i) + (1-\mu_B(x_i))^{\frac{R}{\alpha}})^{\frac{\alpha}{R}}] \\ \quad + \sum_{X_2}[1 - (\mu_A^{\frac{R}{\alpha}}(x_i) + (1-\mu_A(x_i))^{\frac{R}{\alpha}})^{\frac{\alpha}{R}}] \end{array}\right] \end{aligned} \tag{2.13}$$

Adding Eqs. 2.12 and 2.13 gives

$$H_R^\alpha(A \cup B) + H_R^\alpha(A \cap B) = H_R^\alpha(A) + H_R^\alpha(B).$$

Table 2.3 Different values of $H_R^m(A)$ for $R = 0.5$ and for different values of m

α	$\mu_A(x_i)$	0.0	0.1	0.2	0.3	0.4	0.5	0.6	0.7	0.8	0.9	1.0
0.1	$H_{0.5}^{0.1}(A)$	0	0.2469	0.4184	0.5363	0.6058	0.6288	0.6058	0.5363	0.4184	0.2469	0
0.3	$H_{0.5}^{0.3}(A)$	0	0.2797	0.4533	0.5675	0.6331	0.6546	0.6331	0.5675	0.4533	0.2797	0
0.4	$H_{0.5}^{0.4}(A)$	0	0.3004	0.4749	0.5872	0.651	0.6718	0.651	0.5872	0.4749	0.3004	0
0.7	$H_{0.5}^{0.7}(A)$	0	0.3920	0.5700	0.6780	0.7376	0.7568	0.7376	0.6780	0.5700	0.3920	0
0.9	$H_{0.5}^{0.9}(A)$	0	0.5043	0.6915	0.8017	0.8619	0.8811	0.8619	0.8017	0.6915	0.5043	0

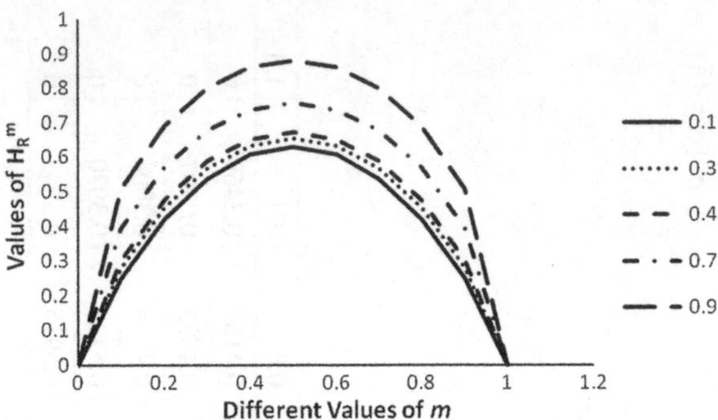

Fig. 2.3 Concave behaviour of $H_R^m(A)$ with respect to m

Hence, 2.2 holds.

In particular,

For $A \in FS(X)$, $A \in FS(X)$ where $\overline{A}$ is the complement of fuzzy set A, it gets

$$H_R^\alpha(A) = H_R^\alpha(\overline{A}) = H_R^\alpha(A \cup \overline{A}) = H_R^\alpha(A \cap \overline{A}) \tag{2.14}$$

Theorem 2.3 *For* $A \in FS(X)$, $H_R^m(A \cup B) + H_R^m(A \cap B) = H_R^m(A) + H_R^m(B)$.

Proof Clearly the result can be proved on similar lines as in Theorem 2.2.

In particular,

For $A \in FS(X)$, $A \in FS(X)$ where $\overline{A}$ is the complement of fuzzy set A, it gets

$$H_R^m(A) = H_R^m(\overline{A}) = H_R^m(A \cup \overline{A}) = H_R^m(A \cap \overline{A}). \tag{2.15}$$

2.3 Numerical Example

Example Let $A = \{(x_i, \mu_A(x_i))/x_i \in X\}$ be a fuzzy set in the universe of discourse $X = \{x_1, x_2, \ldots, x_n\}$. For any real number n, from the operation of power of a fuzzy set:

$$A^n = \{(x_i, [\mu_A(x_i)]^n)/x_i \in X\}.$$

Let us assume a standard fuzzy set A on $X = \{x_1, x_2, \ldots, x_n\}$ defined as:

$$A = (0.2, 0.3, 0.2, 0.4, 0.5)$$

Table 2.4 Different values of $H_R^{\alpha}(A)$ for $m = 0.7$ and for different values of R

R	$\mu_A(x_i)$	0	0.1	0.2	0.3	0.4	0.5	0.6	0.7	0.8	0.9	1
0.1	$H_{0.1}^{0.7}(A)$	0	0.7630	0.9920	1.1244	1.1961	1.219	1.1961	1.1244	0.9920	0.7630	0
0.2	$H_{0.2}^{0.7}(A)$	0	0.6000	0.8000	0.9165	0.9798	1	0.9798	0.9165	0.8000	0.6000	0
0.4	$H_{0.4}^{0.7}(A)$	0	0.4396	0.6207	0.7288	0.7881	0.8071	0.7881	0.7288	0.6207	0.4396	0
0.6	$H_{0.6}^{0.7}(A)$	0	0.3549	0.5313	0.6402	0.7009	0.7205	0.7009	0.6402	0.5313	0.3549	0
0.8	$H_{0.8}^{0.7}(A)$	0	0.3004	0.4749	0.5872	0.6510	0.6718	0.6510	0.5872	0.4749	0.3004	0

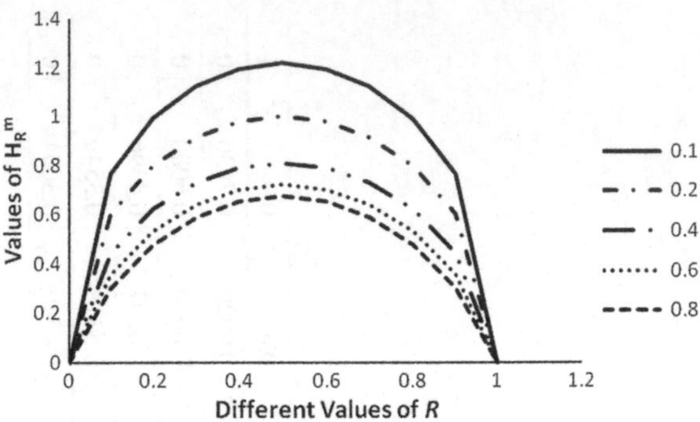

Fig. 2.4 Concave behaviour of $H_R^m(A)$ with respect to R

By taking into consideration the characterization of linguistic variables, A is considered as "LARGE" on X. Using the above operation:

$A^{1/2}$ may be treated as "More or less LARGE";
A^2 may be treated as "Very LARGE";
A^3 may be treated as "Quite very LARGE";
A^4 may be treated as "Very very LARGE"

Next these fuzzy sets are used to compare the above proposed fuzzy entropy measures with one of R-norm fuzzy entropy measure provided in Eq. 2.5. From the point of logical consideration, it may be mentioned that the entropies of fuzzy sets are required to follow the following order pattern:

$$H_R(A^{1/2}) > H_R(A) > H_R(A^2) > H_R(A^3) > H_R(A^4). \tag{2.16}$$

The calculated numerical values of existing fuzzy information measure H_R for these cases are given in Table 2.5.

For any particular value of $\alpha = 0.7$ and different values of R, the calculated numerical values of the proposed measure H_R^α are presented in Table 2.6.

Table 2.7 provides the calculated numerical values of the proposed measure H_R^m, for any particular value of $m = 0.7$ and different values of R.

The numerical values given in Table 2.5 reveal that the R-norm fuzzy entropy measure $H_R(A)$ satisfies Eq. 2.16. The results presented in Tables 2.6 and 2.7

Table 2.5 Numerical values of the R-norm fuzzy entropy measure H_R

H_R	$A^{1/2}$	A	A^2	A^3	A^4
$R = 0.1$	328.9041	258.0100	179.0883	118.0176	79.5181
$R = 0.5$	4.8587	4.4963	2.9554	1.8314	1.1392
$R = 0.9$	4.4643	3.1244	1.7781	0.9322	0.4897
$R = 2$	2.7415	2.3220	1.0367	0.4409	0.1940
$R = 5$	2.3658	1.8854	0.7242	0.2900	0.1242

Table 2.6 Numerical values of the R-norm fuzzy entropy measure H_R^{α}

H_R^{α}	$A^{1/2}$	A	A^2	A^3	A^4
$R = 0.1$	51.2045	47.8754	33.5312	22.6947	15.6592
$R = 0.5$	3.8610	3.5240	2.1409	1.2058	0.6798
$R = 0.9$	3.0597	3.3325	1.3683	0.5984	0.3062
$R = 2$	2.5640	2.1088	0.8703	0.3548	0.1527
$R = 5$	2.2744	1.7958	0.6744	0.2698	0.1156

Table 2.7 Numerical values of R-norm fuzzy entropy measure H_R^m

H_R^m	$A^{1/2}$	A	A^2	A^3	A^4
$R = 0.1$	5.9352	5.5234	3.7551	2.4330	1.5886
$R = 0.5$	3.6502	3.3124	1.9470	1.0613	0.5777
$R = 0.9$	3.1255	2.7618	1.4353	0.6890	0.3335
$R = 2$	2.6660	2.2309	0.9625	0.4010	0.1743
$R = 5$	2.3492	1.8684	0.7144	0.2859	0.1225

clarify that the proposed new generalized R-norm fuzzy entropy measures $H_R^{\alpha}(A)$ and $H_R^m(A)$ satisfy the same:

$$H_R^{\alpha}(A^{1/2}) > H_R^{\alpha}(A) > H_R^{\alpha}(A^2) > H_R^{\alpha}(A^3) > H_R^{\alpha}(A^4) \tag{2.17}$$

$$\text{and } H_R^m(A^{1/2}) > H_R^m(A) > H_R^m(A^2) > H_R^m(A^3) > H_R^m(A^4). \tag{2.18}$$

Thus, the behaviour of new generalized R-norm fuzzy entropy measures $H_R^{\alpha}(A)$ and $H_R^m(A)$ is also consistent for the structured linguistic variables. Table 2.8 and Fig. 2.5 display a declining trend in the numerical values of three entropy measures corresponding to the logical order of fuzzy sets.

Figure 2.5 clearly depicts the similarity between the proposed entropy measures and the existing one. An inequality among two proposed measures of fuzzy information and the existing one may also be seen from Fig. 2.5. That is: $H_R \geq H_R^{\alpha} \geq H_R^m$.

Table 2.8 Calculated numerical values of proposed fuzzy entropy measures and existing one

$R = 0.5$	H_R (Existing)	H_R^{α} (Proposed)	H_R^m (Proposed)
$A^{1/2}$	4.8587	3.8610	3.6502
A	4.4963	3.5240	3.3124
A^2	2.9554	2.1409	1.9470
A^3	1.8314	1.2058	1.0613
A^4	1.1392	0.6798	0.5777

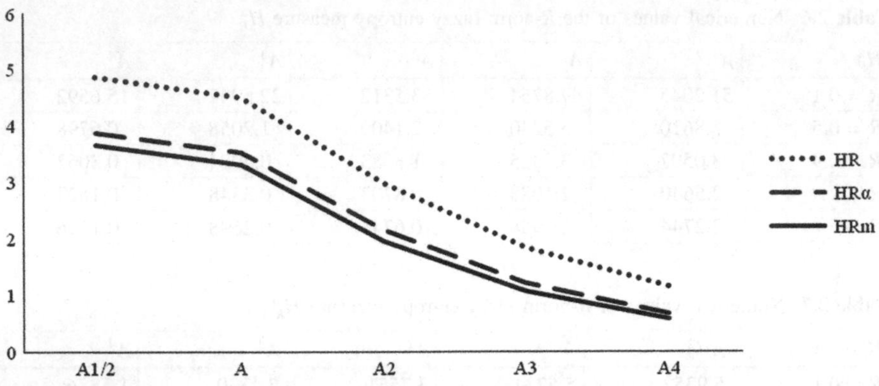

Fig. 2.5 Comparison of numerical values of H_R, H_R^{α} and H_R^{m}

2.4 Generalized Measures of *R*-norm Fuzzy Divergence

As mentioned in Chap. 1, in the recent years, many authors have introduced various measures of divergence for probability distributions and their corresponding divergence measures between fuzzy sets.

Hooda and Sharma [9] introduced the *R*-norm measure of divergence between the probability distribution $P = (p_1, p_2, \ldots, p_n)$ and $Q = (q_1, q_2, \ldots, q_n)$ given by

$$D_R(P,Q) = \frac{R}{R-1}\left[\left(\sum_{i=1}^{n} p_i^R q^{1-R}\right)^{\frac{1}{2}} - 1\right];\ R > 0,\ R \neq 1. \tag{2.19}$$

It is noticed that when $R \rightarrow 1$, the measure (2.19) reduces to Kullback and Leibler [10] divergence measure

$$D(P:Q) = \sum_{i=1}^{n} p_i \log \frac{p_i}{q_i}$$

Analogous to the measure (2.19), Hooda and Bajaj [6] defined the *R*-norm fuzzy divergence measure for fuzzy sets *A* and *B* as

$$I_R(A,B) = \frac{R}{R-1}\sum_{i=1}^{n}\left[\{\mu_A^R(x_i)\mu_B^{1-R}(x_i) + (1-\mu_A(x_i))^R(1-\mu_B(x_i))^{1-R}\}^{\frac{1}{R}} - 1\right]$$
$$R(>0) \neq 1. \tag{2.20}$$

and the corresponding R-norm fuzzy symmetric divergence measure as

$$J_R(A,B) = I_R(A,B) + I_R(B,A) \tag{2.21}$$

2.5 Two New Generalized R-norm Fuzzy Divergence Measures

We now propose the parametric generalized R-norm fuzzy divergence measures corresponding to the measures (2.8) and (2.9) for fuzzy sets A and B of universe of discourse $X = \{x_1, x_2, \ldots, x_n\}$ having the membership values $\mu_A(x_i),\ \mu_B(x_i), i = 1, 2, \ldots, n$, as

$$\begin{aligned} I_R^{\alpha}(A,B) = \frac{R}{R-\alpha}\sum_{i=1}^{n}\left[(\mu_A^{\frac{R}{\alpha}}(x_i)\mu_B^{1-\frac{R}{\alpha}}(x_i) + (1-\mu_A(x_i))^{\frac{R}{\alpha}}(1-\mu_B(x_i))^{1-\frac{R}{\alpha}})^{\frac{\alpha}{R}} - 1\right], \\ 0<\alpha\leq 1,\ R(>0)\neq 1. \end{aligned} \tag{2.22}$$

and

$$\begin{aligned} I_R^{m}(A,B) = \frac{R-m+1}{R-m}\sum_{i=1}^{n}\left[\begin{array}{l}\{\mu_A^{R-m+1}(x_i)\mu_B^{R-m+1}(x_i) \\ \quad +(1-\mu_A(x_i))^{R-m+1}(1-\mu_B(x_i))^{R-m+1}\}^{\frac{1}{R-m+1}} - 1\end{array}\right], \\ R-m+1>0, R\neq m, R, m>0(\neq 1). \end{aligned} \tag{2.23}$$

Theorem 2.4 *The generalized R-norm fuzzy divergence measures* $I_R^{\alpha}(A,B)$ *and* $I_R^{m}(A,B)$ *defined in Eqs. 2.22 and 2.23 are valid measures of fuzzy divergence.*

Proof It is clear from the definitions of measures $I_R^{\alpha}(A,B)$ and $I_R^{m}(A,B)$ in (2.22) and (2.23), respectively, that

(i) $I_R^{\alpha}(A,B)\geq 0$ and $I_R^{m}(A,B)\geq 0$,
(ii) the equality holds if $\mu_A(x_i) = \mu_B(x_i),\ \forall i = 1, 2, \ldots, n$.
(iii) We now prove the convexity of measure $I_R^{\alpha}(A,B)$.

Let us assume that $\sum_{i=1}^{n}\mu_A(x_i) = \alpha_1,\ \sum_{i=1}^{n}\mu_B(x_i) = \beta_1$.

The measure $I_R^\alpha(A,B)$ in (2.22) can be written as

$$I_R^\alpha(A,B) = t_1 \sum_{i=1}^{n} \left[(\mu_A^{\delta_1}(x_i)\mu_B^{1-\delta_1}(x_i) + (1-\mu_A(x_i))^{\delta_1}(1-\mu_B(x_i))^{1-\delta_1})^{\delta_1} - 1 \right]$$

where $t_1 = \frac{R}{R-\alpha}$ and $\left(\delta_1 = \frac{R}{\alpha} > 0\right) \neq 1$.

Then two cases arise.

Case 1: $R<\alpha$

In this case we have $t_1 <0$, $\delta_1 <1$.

$$\left[\sum_{i=1}^{n} \left(\frac{\mu_A(x_i)}{\alpha_1}\right)^{\delta_1} \left(\frac{\mu_B(x_i)}{\beta_1}\right)^{1-\delta_1} -1 \right] \leq 0$$
$$\text{or} \sum_{i=1}^{n} \mu_A^{\delta_1}(x_i)\mu_B^{1-\delta_1}(x_i) \leq \alpha_1^{\delta_1}\beta_1^{1-\delta_1} \tag{2.24}$$

Similarly,

$$\sum_{i=1}^{n} (1-\mu_A(x_i))^{\delta_1}(1-\mu_B(x_i))^{1-\delta_1} \leq (n-\alpha_1)^{\delta_1}(n-\beta_1)^{1-\delta_1} \tag{2.25}$$

Adding Eqs. 2.24 and 2.25

$$\sum_{i=1}^{n} \left(\mu_A^{\delta_1}(x_i)\mu_B^{1-\delta_1}(x_i) + (1-\mu_A(x_i))^{\delta_1}(1-\mu_B(x_i))^{1-\delta_1} \right)$$
$$\leq \left(\alpha_1^{\delta_1}\beta_1^{1-\delta_1} + (n-\alpha_1)^{\delta_1}(n-\beta_1)^{1-\delta_1} \right) \tag{2.26}$$
$$I_R^\alpha(A,B) \geq t_1 \left(\alpha_1^{\delta_1}\beta_1^{1-\delta_1} + (n-\alpha_1)^{\delta_1}(n-\beta_1)^{1-\delta_1} - n \right)$$

Case 2: $R > \alpha$

In this case we have $t_1 > 0$, $\delta_1 > 1$.

$$\left[\sum_{i=1}^{n} \left(\frac{\mu_A(x_i)}{\alpha_1}\right)^{\delta_1} \left(\frac{\mu_B(x_i)}{\beta_1}\right)^{1-\delta_1} -1 \right] \geq 0$$
$$\text{or} \quad \sum_{i=1}^{n} \mu_A^{\delta_1}(x_i)\mu_B^{1-\delta_1}(x_i) \geq \alpha_1^{\delta_1}\beta_1^{1-\delta_1} \tag{2.27}$$

Similarly,

$$\sum_{i=1}^{n} (1-\mu_A(x_i))^{\delta_1}(1-\mu_B(x_i))^{1-\delta_1} \geq (n-\alpha_1)^{\delta_1}(n-\beta_1)^{1-\delta_1} \tag{2.28}$$

Adding Eqs. 2.27 and 2.28

$$\begin{aligned} &\sum_{i=1}^{n}\left(\mu_{_A}^{\delta_1}(x_i)\mu_{_B}^{1-\delta_1}(x_i)+(1-\mu_A(x_i))^{\delta_1}(1-\mu_B(x_i))^{1-\delta_1}\right) \\ &\quad \geq \left(\alpha_1^{\delta_1}\beta_1^{1-\delta_1}+(n-\alpha_1)^{\delta_1}(n-\beta_1)^{1-\delta_1}\right) \\ &\quad I_R^{\alpha}(A,B) \geq t_1\left(\alpha_1^{\delta_1}\beta_1^{1-\delta_1}+(n-\alpha_1)^{\delta_1}(n-\beta_1)^{1-\delta_1}-n\right) \end{aligned} \tag{2.29}$$

Now let $f(\alpha_1) = t\left(\alpha_1^{\delta_1}\beta_1^{1-\delta_1}+(n-\alpha_1)^{\delta_1}(n-\beta_1)^{1-\delta_1}-n\right)$
So that we have

$$f'(\alpha_1) = t_1\left[\delta_1\left(\frac{\alpha_1}{\beta_1}\right)^{\delta_1-1} - \delta_1\left(\frac{n-\alpha_1}{n-\beta_1}\right)^{\delta_1-1}\right] \tag{2.30}$$

$$f''(\alpha_1) = t_1\delta_1(\delta_1-1)\left[\frac{1}{\beta_1}\left(\frac{\alpha_1}{\beta_1}\right)^{\delta_1-2} + \frac{1}{n-\beta_1}\left(\frac{n-\alpha_1}{n-\beta_1}\right)^{\delta_1-2}\right] > 0 \tag{2.31}$$

So this shows that $f(\alpha_1)$ is a convex function of α_1 whose minimum value exists when $\frac{\alpha_1}{\beta_1} = \frac{n-\alpha_1}{n-\beta_1}$, i.e. at $\alpha_1 = \beta_1$ and is equal to zero. So that $f(\alpha_1) > 0$ and vanishes only when $\alpha_1 = \beta_1$.

Hence in view of the definition of fuzzy divergence measure of Bhandari and Pal [2] provided in Sect. 1.3 of Chap. 1, $I_R^{\alpha}(A,B)$ is a valid measure of R-norm fuzzy divergence of order α and consequently it establishes the validity of $J_{_R}^{\alpha}(A,B) = I_{_R}^{\alpha}(A,B) + I_{_R}^{\alpha}(B,A)$ as measure of R-norm symmetric divergence of order α between fuzzy sets A and B.

We now prove the convexity of measure $I_R^m(A,B)$.

Let us assume that $\sum_{i=1}^{n}\mu_A(x_i) = \alpha_2$, $\sum_{i=1}^{n}\mu_B(x_i) = \beta_2$.

The measure $I_R^m(A,B)$ in (2.23) can be written as

$$I_R^m(A,B) = t_2\sum_{i=1}^{n}\left[(\mu_A^{\delta_2}(x_i)\mu_B^{1-\delta_2}(x_i)+(1-\mu_A(x_i))^{\delta_2}(1-\mu_B(x_i))^{1-\delta_2})^{\delta_2}-1\right]$$

where $t_2 = \frac{R-m+1}{R-m}$ and $((\delta_2 = R-m+1) > 0) \neq 1$.

Then two cases arise.

Case 1: $R-m+1<1$

In this case we have $t_2<0$, $\delta_2<1$.

$$\left[\sum_{i=1}^{n}\left(\frac{\mu_A(x_i)}{\alpha_2}\right)^{\delta_2}\left(\frac{\mu_B(x_i)}{\beta_2}\right)^{1-\delta_2}-1\right]\leq 0$$
$$\text{or}\ \sum_{i=1}^{n}\mu_A^{\delta_2}(x_i)\mu_B^{1-\delta_2}(x_i)\leq \alpha_2^{\delta_2}\beta_2^{1-\delta_2} \tag{2.32}$$

Similarly,

$$\sum_{i=1}^{n}(1-\mu_A(x_i))^{\delta_2}(1-\mu_B(x_i))^{1-\delta_2}\leq (n-\alpha_2)^{\delta_2}(n-\beta_2)^{1-\delta_2} \tag{2.33}$$

Adding Eqs. 2.32 and 2.33

$$\sum_{i=1}^{n}\left(\mu_A^{\delta_2}(x_i)\mu_B^{1-\delta_2}(x_i)+(1-\mu_A(x_i))^{\delta_2}(1-\mu_B(x_i))^{1-\delta_2}\right)$$
$$\leq\left(\alpha_2^{\delta_2}\beta_2^{1-\delta_2}+(n-\alpha_2)^{\delta_2}(n-\beta_2)^{1-\delta_2}\right) \tag{2.34}$$
$$I_R^m(A,B)\geq t_2\left(\alpha_2^{\delta_2}\beta_2^{1-\delta_2}+(n-\alpha_2)^{\delta_2}(n-\beta_2)^{1-\delta_2}-n\right)$$

Case 2: $R-m+1>1$

In this case we have $t_2>0$, $\delta_2>1$.

$$\left[\sum_{i=1}^{n}\left(\frac{\mu_A(x_i)}{\alpha_2}\right)^{\delta_2}\left(\frac{\mu_B(x_i)}{\beta_2}\right)^{1-\delta_2}-1\right]\geq 0$$
$$\text{or}\ \sum_{i=1}^{n}\mu_A^{\delta_2}(x_i)\mu_B^{1-\delta_2}(x_i)\geq \alpha_2^{\delta_2}\beta_2^{1-\delta_2} \tag{2.35}$$

Similarly,

$$\sum_{i=1}^{n}(1-\mu_A(x_i))^{\delta_2}(1-\mu_B(x_i))^{1-\delta_2}\geq (n-\alpha_2)^{\delta_2}(n-\beta_2)^{1-\delta_2} \tag{2.36}$$

Adding Eqs. 2.35 and 2.36

$$\sum_{i=1}^{n}\left(\mu_A^{\delta_2}(x_i)\mu_B^{1-\delta_2}(x_i)+(1-\mu_A(x_i))^{\delta_2}(1-\mu_B(x_i))^{1-\delta_2}\right)$$
$$\geq\left(\alpha_2\beta_2^{1-\delta_2}+(n-\alpha_2)^{\delta_2}(n-\beta_2)^{1-\delta_2}\right) \tag{2.37}$$
$$I_R^m(A,B)\geq t_2\left(\alpha_2^{\delta_2}\beta_2^{1-\delta_2}+(n-\alpha_2)^{\delta_2}(n-\beta_2)^{1-\delta_2}-n\right)$$

Now let $g(\alpha_2)=t_2\left(\alpha_2^{\delta_2}\beta_2^{1-\delta_2}+(n-\alpha_2)^{\delta_2}(n-\beta_2)^{1-\delta_2}-n\right)$
So that we have

$$g'(\alpha_2)=t_2\left[\delta_2\left(\frac{\alpha_2}{\beta_2}\right)^{\delta_2-1}-\delta_2\left(\frac{n-\alpha_2}{n-\beta_2}\right)^{\delta_2-1}\right] \tag{2.38}$$

$$g''(\alpha_2)=t_2\delta_2(\delta_2-1)\left[\frac{1}{\beta_2}\left(\frac{\alpha_2}{\beta_2}\right)^{\delta_2-2}+\frac{1}{n-\beta_2}\left(\frac{n-\alpha_2}{n-\beta_2}\right)^{\delta_2-2}\right]>0 \tag{2.39}$$

So this shows that $g(\alpha_2)$ is a convex function of α_2 whose minimum value exists when $\frac{\alpha_2}{\beta_2}=\frac{n-\alpha_2}{n-\beta_2}$, i.e. at $\alpha_2=\beta_2$ and is equal to zero. So that $g(\alpha_2)>0$ and vanishes only when $\alpha_2=\beta_2$.

Hence in view of the definition of fuzzy divergence measure of Bhandari and Pal [2] provided in Sect. 1.3 of Chap. 1, $I_R^m(A,B)$ is a valid measure of R-norm fuzzy divergence of type R and degree m and consequently it establishes the validity of $J_R^m(A,B)=I_R^m(A,B)+I_R^m(B,A)$ as measure of R-norm symmetric divergence of type R and degree m between fuzzy sets A and B.

Limiting and Particular Cases:

(i) when $\alpha\to1$, $m\to1$, measures (2.22) and (2.23) reduce to $I_R(A,B)$.
(ii) when $\alpha\to1$, $m\to1$ with $R=1$, measures (2.22) and (2.23) reduce to Bhandari and Pal [2] fuzzy divergence measure $I(A,B)$.

2.6 Properties of New Generalized *R*-norm Fuzzy Divergence Measures

The proposed generalized R-norm fuzzy divergence measures (2.22) and (2.23) satisfy the following properties.

Theorem 2.5 *For fuzzy sets A and B*

(a) $I_R^\alpha(A\cup B,C)\leq I_R^\alpha(A,C)+I_R^\alpha(B,C)$ *and* $I_R^m(A\cup B,C)\leq I_R^m(A,C)+I_R^m(B,C)$.
(b) $I_R^\alpha(A\cap B,C)\leq I_R^\alpha(A,C)+I_R^\alpha(B,C)$ *and* $I_R^m(A\cap B,C)\leq I_R^m(A,C)+I_R^m(B,C)$.

Proof 2.5(a) Applying Eqs. 2.10 and 2.11

Let us consider the expression

$$
\begin{aligned}
& I_R^{\alpha}(A,C) + I_R^{\alpha}(B,C) - I_R^{\alpha}(A\cup B,C) \\
&= \frac{R}{R-\alpha}\sum_{i=1}^{n}\left[\left(\mu_A^{\frac{R}{\alpha}}(x_i)\mu_C^{1-\frac{R}{\alpha}}(x_i) + (1-\mu_A(x_i))^{\frac{R}{\alpha}}(1-\mu_C(x_i))^{1-\frac{R}{\alpha}}\right)^{\frac{\alpha}{R}} - 1\right] \\
&+ \frac{R}{R-\alpha}\sum_{i=1}^{n}\left[\left(\mu_B^{\frac{R}{\alpha}}(x_i)\mu_C^{1-\frac{R}{\alpha}}(x_i) + (1-\mu_B(x_i))^{\frac{R}{\alpha}}(1-\mu_C(x_i))^{1-\frac{R}{\alpha}}\right)^{\frac{\alpha}{R}} - 1\right] \\
&- \frac{R}{R-\alpha}\sum_{i=1}^{n}\left[\left(\mu_{A\cup B}^{\frac{R}{\alpha}}(x_i)\mu_C^{1-\frac{R}{\alpha}}(x_i) + (1-\mu_{A\cup B}(x_i))^{\frac{R}{\alpha}}(1-\mu_C(x_i))^{1-\frac{R}{\alpha}}\right)^{\frac{\alpha}{R}} - 1\right] \\
&= \frac{R}{R-\alpha}\sum_{X_1}\left[\left(\mu_B^{\frac{R}{\alpha}}(x_i)\mu_C^{1-\frac{R}{\alpha}}(x_i) + (1-\mu_B(x_i))^{\frac{R}{\alpha}}(1-\mu_C(x_i))^{1-\frac{R}{\alpha}}\right)^{\frac{\alpha}{R}} - 1\right] \\
&+ \frac{R}{R-\alpha}\sum_{X_2}\left[\left(\mu_A^{\frac{R}{\alpha}}(x_i)\mu_C^{1-\frac{R}{\alpha}}(x_i) + (1-\mu_A(x_i))^{\frac{R}{\alpha}}(1-\mu_C(x_i))^{1-\frac{R}{\alpha}}\right)^{\frac{\alpha}{R}} - 1\right] \geq 0.
\end{aligned}
$$

Now, let us consider the expression

$$
\begin{aligned}
& I_R^{m}(A,C) + I_R^{m}(B,C) - I_R^{m}(A\cup B,C) \\
&= \frac{R-m+1}{R-m}\sum_{i=1}^{n}\left[\{\mu_A^{R-m+1}(x_i)\mu_C^{R-m+1}(x_i) + (1-\mu_A(x_i))^{R-m+1}(1-\mu_C(x_i))^{R-m+1}\}^{\frac{1}{R-m+1}} - 1\right] \\
&+ \frac{R-m+1}{R-m}\sum_{i=1}^{n}\left[\{\mu_B^{R-m+1}(x_i)\mu_C^{R-m+1}(x_i) + (1-\mu_B(x_i))^{R-m+1}(1-\mu_C(x_i))^{R-m+1}\}^{\frac{1}{R-m+1}} - 1\right] \\
&- \frac{R-m+1}{R-m}\sum_{i=1}^{n}\left[\{\mu_{A\cup B}^{R-m+1}(x_i)\mu_C^{R-m+1}(x_i) + (1-\mu_{A\cup B}(x_i))^{R-m+1}(1-\mu_C(x_i))^{R-m+1}\}^{\frac{1}{R-m+1}} - 1\right] \\
&= \frac{R-m+1}{R-m}\sum_{X_1}\left[\{\mu_B^{R-m+1}(x_i)\mu_C^{R-m+1}(x_i) + (1-\mu_B(x_i))^{R-m+1}(1-\mu_C(x_i))^{R-m+1}\}^{\frac{1}{R-m+1}} - 1\right] \\
&+ \frac{R-m+1}{R-m}\sum_{X_2}\left[\{\mu_A^{R-m+1}(x_i)\mu_C^{R-m+1}(x_i) + (1-\mu_A(x_i))^{R-m+1}(1-\mu_C(x_i))^{R-m+1}\}^{\frac{1}{R-m+1}} - 1\right] \geq 0.
\end{aligned}
$$

Hence, 2.5(a) holds.

Similarly, 2.5(b) can also be proved.

Theorem 2.6 *For fuzzy sets A and B*

(a) $I_R^{\alpha}(A\cup B,C) + I_R^{\alpha}(A\cap B,C) = I_R^{\alpha}(A,C) + I_R^{\alpha}(B,C)$.
(b) $I_R^{m}(A\cup B,C) + I_R^{m}(A\cap B,C) = I_R^{m}(A,C) + I_R^{m}(B,C)$.

Proof 2.6(a) Applying Eqs. 2.10 and 2.11 we get

$$
\begin{aligned}
&I_R^{\alpha}(A\cup B, C)\\
&= \frac{R}{R-\alpha}\sum_{i=1}^{n}\left[(\mu_{A\cup B}^{\frac{R}{\alpha}}(x_i)\mu_C^{1-\frac{R}{\alpha}}(x_i) + (1-\mu_{A\cup B}(x_i))^{\frac{R}{\alpha}}(1-\mu_C(x_i))^{1-\frac{R}{\alpha}})^{\frac{\alpha}{R}} - 1\right]\\
&= \sum_{X_1}\left[(\mu_A^{\frac{R}{\alpha}}(x_i)\mu_C^{1-\frac{R}{\alpha}}(x_i) + (1-\mu_A(x_i))^{\frac{R}{\alpha}}(1-\mu_C(x_i))^{1-\frac{R}{\alpha}})^{\frac{\alpha}{R}} - 1\right]\\
&\quad + \sum_{X_2}\left[(\mu_B^{\frac{R}{\alpha}}(x_i)\mu_C^{1-\frac{R}{\alpha}}(x_i) + (1-\mu_B(x_i))^{\frac{R}{\alpha}}(1-\mu_C(x_i))^{1-\frac{R}{\alpha}})^{\frac{\alpha}{R}} - 1\right]
\end{aligned}
\tag{2.40}
$$

$$
\begin{aligned}
&I_R^{\alpha}(A\cap B, C)\\
&= \frac{R}{R-\alpha}\sum_{i=1}^{n}\left[(\mu_{A\cap B}^{\frac{R}{\alpha}}(x_i)\mu_C^{1-\frac{R}{\alpha}}(x_i) + (1-\mu_{A\cap B}(x_i))^{\frac{R}{\alpha}}(1-\mu_C(x_i))^{1-\frac{R}{\alpha}})^{\frac{\alpha}{R}} - 1\right]\\
&= \sum_{X_1}\left[(\mu_B^{\frac{R}{\alpha}}(x_i)\mu_C^{1-\frac{R}{\alpha}}(x_i) + (1-\mu_B(x_i))^{\frac{R}{\alpha}}(1-\mu_C(x_i))^{1-\frac{R}{\alpha}})^{\frac{\alpha}{R}} - 1\right]\\
&\quad + \sum_{X_2}\left[(\mu_A^{\frac{R}{\alpha}}(x_i)\mu_C^{1-\frac{R}{\alpha}}(x_i) + (1-\mu_A(x_i))^{\frac{R}{\alpha}}(1-\mu_C(x_i))^{1-\frac{R}{\alpha}})^{\frac{\alpha}{R}} - 1\right]
\end{aligned}
\tag{2.41}
$$

On adding Eqs. 2.40 and 2.41 we get the result.
Hence, 2.6(a) holds.
2.6(b)

$$
\begin{aligned}
&I_R^{m}(A\cup B, C)\\
&= \frac{R-m+1}{R-m}\sum_{i=1}^{n}\left[\begin{array}{l}\{\mu_{A\cup B}^{R-m+1}(x_i)\mu_C^{R-m+1}(x_i)\\ \quad + (1-\mu_{A\cup B}(x_i))^{R-m+1}(1-\mu_C(x_i))^{R-m+1}\}^{\frac{1}{R-m+1}} - 1\end{array}\right]\\
&= \frac{R-m+1}{R-m}\sum_{X_1}\left[\begin{array}{l}\{\mu_A^{R-m+1}(x_i)\mu_C^{R-m+1}(x_i)\\ \quad + (1-\mu_A(x_i))^{R-m+1}(1-\mu_C(x_i))^{R-m+1}\}^{\frac{1}{R-m+1}} - 1\end{array}\right]\\
&\quad + \frac{R-m+1}{R-m}\sum_{X_2}\left[\begin{array}{l}\{\mu_B^{R-m+1}(x_i)\mu_C^{R-m+1}(x_i)\\ \quad + (1-\mu_B(x_i))^{R-m+1}(1-\mu_C(x_i))^{R-m+1}\}^{\frac{1}{R-m+1}} - 1\end{array}\right]
\end{aligned}
\tag{2.42}
$$

$$I_R^m(A\cap B, C)$$
$$=\frac{R-m+1}{R-m}\sum_{i=1}^{n}\left[\begin{array}{l}\{\mu_{A\cap B}^{R-m+1}(x_i)\mu_C^{R-m+1}(x_i)\\ \quad+(1-\mu_{A\cap B}(x_i))^{R-m+1}(1-\mu_C(x_i))^{R-m+1}\}^{\frac{1}{R-m+1}}-1\end{array}\right]$$
$$=\frac{R-m+1}{R-m}\sum_{X_1}\left[\begin{array}{l}\{\mu_B^{R-m+1}(x_i)\mu_C^{R-m+1}(x_i)\\ \quad+(1-\mu_B(x_i))^{R-m+1}(1-\mu_C(x_i))^{R-m+1}\}^{\frac{1}{R-m+1}}-1\end{array}\right]$$
$$+\frac{R-m+1}{R-m}\sum_{X_2}\left[\begin{array}{l}\{\mu_A^{R-m+1}(x_i)\mu_C^{R-m+1}(x_i)\\ \quad+(1-\mu_A(x_i))^{R-m+1}(1-\mu_C(x_i))^{R-m+1}\}^{\frac{1}{R-m+1}}-1\end{array}\right] \tag{2.43}$$

On adding Eqs. 2.42 and 2.43 we get the result.
Hence, 2.6(b) holds.

Theorem 2.7 *For fuzzy sets A and B*

(a) $I_R^\alpha(A\cup B, A\cap B)+I_R^\alpha(A\cap B, A\cup B)=I_R^\alpha(A,B)+I_R^\alpha(B,A)$.
(b) $I_R^m(A\cup B, A\cap B)+I_R^m(A\cap B, A\cup B)=I_R^m(A,B)+I_R^m(B,A)$.

Proof 2.7(a) Applying Eqs. 2.10 and 2.11 we get

$$I_R^\alpha(A\cup B, A\cap B)$$
$$=\frac{R}{R-\alpha}\sum_{i=1}^{n}\left[(\mu_{A\cup B}^{\frac{R}{\alpha}}(x_i)\mu_{A\cap B}^{1-\frac{R}{\alpha}}(x_i)+(1-\mu_{A\cup B}(x_i))^{\frac{R}{\alpha}}(1-\mu_{A\cap B}(x_i))^{1-\frac{R}{\alpha}})^{\frac{\alpha}{R}}-1\right]$$
$$=\frac{R}{R-\alpha}\sum_{X_1}\left[(\mu_A^{\frac{R}{\alpha}}(x_i)\mu_B^{1-\frac{R}{\alpha}}(x_i)+(1-\mu_A(x_i))^{\frac{R}{\alpha}}(1-\mu_B(x_i))^{1-\frac{R}{\alpha}})^{\frac{\alpha}{R}}-1\right]$$
$$+\frac{R}{R-\alpha}\sum_{X_2}\left[(\mu_B^{\frac{R}{\alpha}}(x_i)\mu_A^{1-\frac{R}{\alpha}}(x_i)+(1-\mu_B(x_i))^{\frac{R}{\alpha}}(1-\mu_A(x_i))^{1-\frac{R}{\alpha}})^{\frac{\alpha}{R}}-1\right] \tag{2.44}$$

Now

$$I_R^\alpha(A\cap B, A\cup B)$$
$$=\frac{R}{R-\alpha}\sum_{i=1}^{n}\left[(\mu_{A\cap B}^{\frac{R}{\alpha}}(x_i)\mu_{A\cup B}^{1-\frac{R}{\alpha}}(x_i)+(1-\mu_{A\cap B}(x_i))^{\frac{R}{\alpha}}(1-\mu_{A\cup B}(x_i))^{1-\frac{R}{\alpha}})^{\frac{\alpha}{R}}-1\right]$$
$$=\frac{R}{R-\alpha}\sum_{X_1}\left[(\mu_B^{\frac{R}{\alpha}}(x_i)\mu_A^{1-\frac{R}{\alpha}}(x_i)+(1-\mu_B(x_i))^{\frac{R}{\alpha}}(1-\mu_A(x_i))^{1-\frac{R}{\alpha}})^{\frac{\alpha}{R}}-1\right]$$
$$+\frac{R}{R-\alpha}\sum_{X_2}\left[(\mu_A^{\frac{R}{\alpha}}(x_i)\mu_B^{1-\frac{R}{\alpha}}(x_i)+(1-\mu_A(x_i))^{\frac{R}{\alpha}}(1-\mu_B(x_i))^{1-\frac{R}{\alpha}})^{\frac{\alpha}{R}}-1\right] \tag{2.45}$$

Adding Eqs. 2.44 and 2.45 we get

$$
\begin{aligned}
&I_R^\alpha(A\cup B, A\cap B) + I_R^\alpha(A\cap B, A\cup B)\\
&= \frac{R}{R-\alpha}\sum_{i=1}^{n}\left[(\mu_A^{\frac{R}{\alpha}}(x_i)\mu_B^{1-\frac{R}{\alpha}}(x_i) + (1-\mu_A(x_i))^{\frac{R}{\alpha}}(1-\mu_B(x_i))^{1-\frac{R}{\alpha}})^{\frac{\alpha}{R}} - 1\right]\\
&\quad + \frac{R}{R-\alpha}\sum_{i=1}^{n}\left[(\mu_B^{\frac{R}{\alpha}}(x_i)\mu_A^{1-\frac{R}{\alpha}}(x_i) + (1-\mu_B(x_i))^{\frac{R}{\alpha}}(1-\mu_A(x_i))^{1-\frac{R}{\alpha}})^{\frac{\alpha}{R}} - 1\right]\\
&= I_R^\alpha(A,B) + I_R^\alpha(B,A)
\end{aligned}
$$

Hence, 2.7(a) holds.
2.7(b)

$$
\begin{aligned}
&I_R^m(A\cup B, A\cap B)\\
&= \frac{R-m+1}{R-m}\sum_{i=1}^{n}\left[\begin{array}{l}\{\mu_{A\cup B}^{R-m+1}(x_i)\mu_{A\cap B}^{R-m+1}(x_i)\\ \quad + (1-\mu_{A\cup B}(x_i))^{R-m+1}(1-\mu_{A\cap B}(x_i))^{R-m+1}\}^{\frac{1}{R-m+1}} - 1\end{array}\right]\\
&= \frac{R-m+1}{R-m}\sum_{X_1}\left[\begin{array}{l}\{\mu_A^{R-m+1}(x_i)\mu_B^{R-m+1}(x_i)\\ \quad + (1-\mu_A(x_i))^{R-m+1}(1-\mu_B(x_i))^{R-m+1}\}^{\frac{1}{R-m+1}} - 1\end{array}\right]\\
&\quad + \frac{R-m+1}{R-m}\sum_{X_2}\left[\begin{array}{l}\{\mu_B^{R-m+1}(x_i)\mu_A^{R-m+1}(x_i)\\ \quad + (1-\mu_B(x_i))^{R-m+1}(1-\mu_A(x_i))^{R-m+1}\}^{\frac{1}{R-m+1}} - 1\end{array}\right]
\end{aligned}
\tag{2.46}
$$

Now

$$
\begin{aligned}
&I_R^m(A\cap B, A\cup B)\\
&= \frac{R-m+1}{R-m}\sum_{i=1}^{n}\left[\begin{array}{l}\{\mu_{A\cap B}^{R-m+1}(x_i)\mu_{A\cup B}^{R-m+1}(x_i)\\ \quad + (1-\mu_{A\cap B}(x_i))^{R-m+1}(1-\mu_{A\cup B}(x_i))^{R-m+1}\}^{\frac{1}{R-m+1}} - 1\end{array}\right]\\
&= \frac{R-m+1}{R-m}\sum_{X_1}\left[\begin{array}{l}\{\mu_B^{R-m+1}(x_i)\mu_A^{R-m+1}(x_i)\\ \quad + (1-\mu_B(x_i))^{R-m+1}(1-\mu_A(x_i))^{R-m+1}\}^{\frac{1}{R-m+1}} - 1\end{array}\right]\\
&\quad + \frac{R-m+1}{R-m}\sum_{X_2}\left[\begin{array}{l}\{\mu_A^{R-m+1}(x_i)\mu_B^{R-m+1}(x_i)\\ \quad + (1-\mu_A(x_i))^{R-m+1}(1-\mu_B(x_i))^{R-m+1}\}^{\frac{1}{R-m+1}} - 1\end{array}\right]
\end{aligned}
\tag{2.47}
$$

Adding Eqs. 2.46 and 2.47 we get

$$\begin{aligned}
&I_R^m(A\cup B, A\cap B) + I_R^m(A\cap B, A\cup B)\\
&= \frac{R-m+1}{R-m}\sum_{i=1}^{n}\left[\{\mu_A^{R-m+1}(x_i)\mu_B^{R-m+1}(x_i) + (1-\mu_A(x_i))^{R-m+1}(1-\mu_B(x_i))^{R-m+1}\}^{\frac{1}{R-m+1}} - 1\right]\\
&\quad + \frac{R-m+1}{R-m}\sum_{i=1}^{n}\left[\{\mu_B^{R-m+1}(x_i)\mu_A^{R-m+1}(x_i) + (1-\mu_B(x_i))^{R-m+1}(1-\mu_A(x_i))^{R-m+1}\}^{\frac{1}{R-m+1}} - 1\right]\\
&= I_R^m(A,B) + I_R^m(B,A)
\end{aligned}$$

Hence, 2.7(b) holds.

Theorem 2.8 *For fuzzy sets A and B*

(a) $I_R^\alpha(A, A\cup B) + I_R^\alpha(A, A\cap B) = I_R^\alpha(A,B)$;
$I_R^m(A, A\cup B) + I_R^m(A, A\cap B) = I_R^m(A,B)$.

(b) $I_R^\alpha(B, A\cup B) + I_R^\alpha(B, A\cap B) = I_R^\alpha(B,A)$;
$I_R^m(B, A\cup B) + I_R^m(B, A\cap B) = I_R^m(B,A)$.

Proof 2.8(a) Applying Eqs. 2.10 and 2.11 we get

$$\begin{aligned}
&I_R^\alpha(A, A\cup B)\\
&= \frac{R}{R-\alpha}\sum_{i=1}^{n}\left[(\mu_A^{\frac{R}{\alpha}}(x_i)\mu_{A\cup B}^{1-\frac{R}{\alpha}}(x_i) + (1-\mu_A(x_i))^{\frac{R}{\alpha}}(1-\mu_{A\cup B}(x_i))^{1-\frac{R}{\alpha}})^{\frac{\alpha}{R}} - 1\right]\\
&= \frac{R}{R-\alpha}\sum_{X_1}\left[(\mu_A^{\frac{R}{\alpha}}(x_i)\mu_A^{1-\frac{R}{\alpha}}(x_i) + (1-\mu_A(x_i))^{\frac{R}{\alpha}}(1-\mu_A(x_i))^{1-\frac{R}{\alpha}})^{\frac{\alpha}{R}} - 1\right]\\
&\quad + \frac{R}{R-\alpha}\sum_{X_2}\left[(\mu_A^{\frac{R}{\alpha}}(x_i)\mu_B^{1-\frac{R}{\alpha}}(x_i) + (1-\mu_A(x_i))^{\frac{R}{\alpha}}(1-\mu_B(x_i))^{1-\frac{R}{\alpha}})^{\frac{\alpha}{R}} - 1\right]\\
&= \frac{R}{R-\alpha}\sum_{X_2}\left[(\mu_A^{\frac{R}{\alpha}}(x_i)\mu_B^{1-\frac{R}{\alpha}}(x_i) + (1-\mu_A(x_i))^{\frac{R}{\alpha}}(1-\mu_B(x_i))^{1-\frac{R}{\alpha}})^{\frac{\alpha}{R}} - 1\right]
\end{aligned} \tag{2.48}$$

Now

$$\begin{aligned}
&I_R^\alpha(A, A\cap B)\\
&= \frac{R}{R-\alpha}\sum_{X_1}\left[(\mu_A^{\frac{R}{\alpha}}(x_i)\mu_B^{1-\frac{R}{\alpha}}(x_i) + (1-\mu_A(x_i))^{\frac{R}{\alpha}}(1-\mu_B(x_i))^{1-\frac{R}{\alpha}})^{\frac{\alpha}{R}} - 1\right]\\
&\quad + \frac{R}{R-\alpha}\sum_{X_2}\left[(\mu_A^{\frac{R}{\alpha}}(x_i)\mu_A^{1-\frac{R}{\alpha}}(x_i) + (1-\mu_A(x_i))^{\frac{R}{\alpha}}(1-\mu_A(x_i))^{1-\frac{R}{\alpha}})^{\frac{\alpha}{R}} - 1\right]\\
&= \frac{R}{R-\alpha}\sum_{X_1}\left[(\mu_A^{\frac{R}{\alpha}}(x_i)\mu_B^{1-\frac{R}{\alpha}}(x_i) + (1-\mu_A(x_i))^{\frac{R}{\alpha}}(1-\mu_B(x_i))^{1-\frac{R}{\alpha}})^{\frac{\alpha}{R}} - 1\right]
\end{aligned} \tag{2.49}$$

Adding Eqs. 2.48 and 2.49 we get

$$
\begin{aligned}
& I_R^{\alpha}(A, A \cup B) + I_R^{\alpha}(A, A \cap B) \\
&= \frac{R}{R-\alpha} \sum_{X_1} \left[(\mu_A^{\frac{R}{\alpha}}(x_i) \mu_B^{1-\frac{R}{\alpha}}(x_i) + (1-\mu_A(x_i))^{\frac{R}{\alpha}} (1-\mu_B(x_i))^{1-\frac{R}{\alpha}})^{\frac{\alpha}{R}} - 1 \right] \\
&= \frac{R}{R-\alpha} \sum_{X_2} \left[(\mu_A^{\frac{R}{\alpha}}(x_i) \mu_B^{1-\frac{R}{\alpha}}(x_i) + (1-\mu_A(x_i))^{\frac{R}{\alpha}} (1-\mu_B(x_i))^{1-\frac{R}{\alpha}})^{\frac{\alpha}{R}} - 1 \right] \\
&= \frac{R}{R-\alpha} \sum_{i=1}^{n} \left[(\mu_A^{\frac{R}{\alpha}}(x_i) \mu_B^{1-\frac{R}{\alpha}}(x_i) + (1-\mu_A(x_i))^{\frac{R}{\alpha}} (1-\mu_B(x_i))^{1-\frac{R}{\alpha}})^{\frac{\alpha}{R}} - 1 \right] \\
&= I_R^{\alpha}(A, B)
\end{aligned}
$$

Hence, 2.8(a) holds for $I_R^{\alpha}(A, B)$ and similar result holds for $I_R^{m}(A, B)$.
2.8(b)

$$
\begin{aligned}
& I_R^{\alpha}(B, A \cup B) \\
&= \frac{R}{R-\alpha} \sum_{i=1}^{n} \left[(\mu_B^{\frac{R}{\alpha}}(x_i) \mu_{A \cup B}^{1-\frac{R}{\alpha}}(x_i) + (1-\mu_B(x_i))^{\frac{R}{\alpha}} (1-\mu_{A \cup B}(x_i))^{1-\frac{R}{\alpha}})^{\frac{\alpha}{R}} - 1 \right] \\
&= \frac{R}{R-\alpha} \sum_{X_1} \left[(\mu_B^{\frac{R}{\alpha}}(x_i) \mu_A^{1-\frac{R}{\alpha}}(x_i) + (1-\mu_B(x_i))^{\frac{R}{\alpha}} (1-\mu_A(x_i))^{1-\frac{R}{\alpha}})^{\frac{\alpha}{R}} - 1 \right] \\
&\quad + \frac{R}{R-\alpha} \sum_{X_2} \left[(\mu_B^{\frac{R}{\alpha}}(x_i) \mu_B^{1-\frac{R}{\alpha}}(x_i) + (1-\mu_B(x_i))^{\frac{R}{\alpha}} (1-\mu_B(x_i))^{1-\frac{R}{\alpha}})^{\frac{\alpha}{R}} - 1 \right] \\
&= \frac{R}{R-\alpha} \sum_{X_1} \left[(\mu_B^{\frac{R}{\alpha}}(x_i) \mu_A^{1-\frac{R}{\alpha}}(x_i) + (1-\mu_B(x_i))^{\frac{R}{\alpha}} (1-\mu_A(x_i))^{1-\frac{R}{\alpha}})^{\frac{\alpha}{R}} - 1 \right]
\end{aligned}
\tag{2.50}
$$

Now

$$
\begin{aligned}
& I_R^{\alpha}(B, A \cap B) \\
&= \frac{R}{R-\alpha} \sum_{i=1}^{n} \left[(\mu_B^{\frac{R}{\alpha}}(x_i) \mu_{A \cap B}^{1-\frac{R}{\alpha}}(x_i) + (1-\mu_B(x_i))^{\frac{R}{\alpha}} (1-\mu_{A \cap B}(x_i))^{1-\frac{R}{\alpha}})^{\frac{\alpha}{R}} - 1 \right] \\
&= \frac{R}{R-\alpha} \sum_{X_1} \left[(\mu_B^{\frac{R}{\alpha}}(x_i) \mu_B^{1-\frac{R}{\alpha}}(x_i) + (1-\mu_B(x_i))^{\frac{R}{\alpha}} (1-\mu_B(x_i))^{1-\frac{R}{\alpha}})^{\frac{\alpha}{R}} - 1 \right] \\
&\quad + \frac{R}{R-\alpha} \sum_{X_2} \left[(\mu_B^{\frac{R}{\alpha}}(x_i) \mu_A^{1-\frac{R}{\alpha}}(x_i) + (1-\mu_B(x_i))^{\frac{R}{\alpha}} (1-\mu_A(x_i))^{1-\frac{R}{\alpha}})^{\frac{\alpha}{R}} - 1 \right] \\
&= \frac{R}{R-\alpha} \sum_{X_2} \left[(\mu_B^{\frac{R}{\alpha}}(x_i) \mu_A^{1-\frac{R}{\alpha}}(x_i) + (1-\mu_B(x_i))^{\frac{R}{\alpha}} (1-\mu_A(x_i))^{1-\frac{R}{\alpha}})^{\frac{\alpha}{R}} - 1 \right]
\end{aligned}
\tag{2.51}
$$

Adding Eqs. 2.50 and 2.51 we get

$$
\begin{aligned}
& I_R^\alpha(B, A\cup B) + I_R^\alpha(B, A\cap B) \\
&= \frac{R}{R-\alpha}\sum_{X_1}\left[(\mu_B^{\frac{R}{\alpha}}(x_i)\mu_A^{1-\frac{R}{\alpha}}(x_i) + (1-\mu_B(x_i))^{\frac{R}{\alpha}}(1-\mu_A(x_i))^{1-\frac{R}{\alpha}})^{\frac{\alpha}{R}} - 1\right] \\
&= \frac{R}{R-\alpha}\sum_{X_2}\left[(\mu_B^{\frac{R}{\alpha}}(x_i)\mu_A^{1-\frac{R}{\alpha}}(x_i) + (1-\mu_B(x_i))^{\frac{R}{\alpha}}(1-\mu_A(x_i))^{1-\frac{R}{\alpha}})^{\frac{\alpha}{R}} - 1\right] \\
&= \frac{R}{R-\alpha}\sum_{i=1}^{n}\left[(\mu_B^{\frac{R}{\alpha}}(x_i)\mu_A^{1-\frac{R}{\alpha}}(x_i) + (1-\mu_B(x_i))^{\frac{R}{\alpha}}(1-\mu_A(x_i))^{1-\frac{R}{\alpha}})^{\frac{\alpha}{R}} - 1\right] \\
&= I_R^\alpha(B,A)
\end{aligned}
$$

Hence, 2.8(b) holds for $I_R^\alpha(A,B)$ and similar result holds for $I_R^m(A,B)$.

Theorem 2.9 *For fuzzy sets A and B*

(a) $I_R^\alpha(A,B) = I_R^\alpha(A^c,B^c)$; $I_R^m(A,B) = I_R^m(A^c,B^c)$.
(b) $I_R^\alpha(A^c,B) = I_R^\alpha(A,B^c)$; $I_R^m(A^c,B) = I_R^m(A,B^c)$.
(c) $I_R^\alpha(A,B) + I_R^\alpha(A^c,B) = I_R^\alpha(A^c,B^c) + I_R^\alpha(A^c,B)$;

$$I_R^m(A,B) + I_R^m(A^c,B) = I_R^m(A^c,B^c) + I_R^m(A^c,B).$$

Proof 2.9(a)

$$
\begin{aligned}
& I_R^\alpha(A^c,B^c) \\
&= \frac{R}{R-\alpha}\sum_{i=1}^{n}\left[(\mu_{A^c}^{\frac{R}{\alpha}}(x_i)\mu_{B^c}^{1-\frac{R}{\alpha}}(x_i) + (1-\mu_{A^c}(x_i))^{\frac{R}{\alpha}}(1-\mu_{B^c}(x_i))^{1-\frac{R}{\alpha}})^{\frac{\alpha}{R}} - 1\right] \\
&= \frac{R}{R-\alpha}\sum_{i=1}^{n}\left[((1-\mu_A(x_i))^{\frac{R}{\alpha}}(1-\mu_B(x_i))^{1-\frac{R}{\alpha}} + \mu_A^{\frac{R}{\alpha}}(x_i)\mu_B^{1-\frac{R}{\alpha}}(x_i))^{\frac{\alpha}{R}} - 1\right] \\
&= I_R^\alpha(A,B)
\end{aligned}
$$

Now

$$
\begin{aligned}
& I_R^m(A^c,B^c) \\
&= \frac{R-m+1}{R-m}\sum_{i=1}^{n}\left[\begin{array}{l}\{\mu_{A^c}^{R-m+1}(x_i)\mu_{B^c}^{R-m+1}(x_i) \\ \quad + (1-\mu_{A^c}(x_i))^{R-m+1}(1-\mu_{B^c}(x_i))^{R-m+1}\}^{\frac{1}{R-m+1}} - 1\end{array}\right] \\
&= \frac{R-m+1}{R-m}\sum_{i=1}^{n}\left[\begin{array}{l}\{(1-\mu_A(x_i))^{R-m+1}(1-\mu_B(x_i))^{R-m+1} \\ \quad + \mu_A^{R-m+1}(x_i)\mu_B^{R-m+1}(x_i)\}^{\frac{1}{R-m+1}} - 1\end{array}\right] = I_R^m(A,B)
\end{aligned}
$$

Hence, 2.9(a) holds.

2.9(b)

$$\begin{aligned} I_R^\alpha(A^c, B) &= \frac{R}{R-\alpha}\sum_{i=1}^{n}\left[(\mu_{A^c}^{\frac{R}{\alpha}}(x_i)\mu_B^{1-\frac{R}{\alpha}}(x_i) + (1-\mu_{A^c}(x_i))^{\frac{R}{\alpha}}(1-\mu_B(x_i))^{1-\frac{R}{\alpha}})^{\frac{\alpha}{R}} - 1\right] \\ &= \frac{R}{R-\alpha}\sum_{i=1}^{n}\left[((1-\mu_A(x_i))^{\frac{R}{\alpha}}\mu_B^{1-\frac{R}{\alpha}}(x_i) + \mu_A^{\frac{R}{\alpha}}(x_i)(1-\mu_B(x_i))^{1-\frac{R}{\alpha}})^{\frac{\alpha}{R}} - 1\right] \end{aligned} \tag{2.52}$$

$$\begin{aligned} I_R^\alpha(A, B^c) &= \frac{R}{R-\alpha}\sum_{i=1}^{n}\left[(\mu_A^{\frac{R}{\alpha}}(x_i)\mu_{B^c}^{1-\frac{R}{\alpha}}(x_i) + (1-\mu_A(x_i))^{\frac{R}{\alpha}}(1-\mu_{B^c}(x_i))^{1-\frac{R}{\alpha}})^{\frac{\alpha}{R}} - 1\right] \\ &= \frac{R}{R-\alpha}\sum_{i=1}^{n}\left[(\mu_A^{\frac{R}{\alpha}}(x_i)(1-\mu_B(x_i))^{1-\frac{R}{\alpha}} + (1-\mu_A(x_i))^{\frac{R}{\alpha}}\mu_B^{1-\frac{R}{\alpha}}(x_i))^{\frac{\alpha}{R}} - 1\right] \end{aligned} \tag{2.53}$$

From Eqs. 2.52 and 2.53 we get $I_R^\alpha(A^c, B) = I_R^\alpha(A, B^c)$.
Hence, 2.9(b) holds for $I_R^\alpha(A, B)$ and similar result holds for $I_R^m(A, B)$.
2.9(c) It obviously follows from 2.9(a) and 2.9(b).

2.7 Relation Between Proposed Generalized R-norm Fuzzy Entropy and Divergence Measures

Theorem 2.10 *For fuzzy sets A and B, the relation between* $H_R^\alpha(A)$ *and* $I_R^\alpha(A, B)$ *is given by*

$$I_R^\alpha\left(A, \left[\frac{1}{2}\right]\right) = \left[\left(\frac{1}{2}\right)^{\frac{\alpha-R}{R}} - 1\right]\frac{nR}{R-\alpha} - \left(\frac{1}{2}\right)^{\frac{\alpha-R}{R}} H_R^\alpha(A).$$

Proof We know from Eq. 2.22 that

$$I_R^\alpha(A, B) = \frac{R}{R-\alpha}\sum_{i=1}^{n}\left[(\mu_A^{\frac{R}{\alpha}}(x_i)\mu_B^{1-\frac{R}{\alpha}}(x_i) + (1-\mu_A(x_i))^{\frac{R}{\alpha}}(1-\mu_B(x_i))^{1-\frac{R}{\alpha}})^{\frac{\alpha}{R}} - 1\right]$$

$$\begin{aligned} I_R^\alpha\left(A, \left[\frac{1}{2}\right]\right) &= \frac{R}{R-\alpha}\sum_{i=1}^{n}\left[(\mu_A^{\frac{R}{\alpha}}(x_i)\left(\frac{1}{2}\right)^{1-\frac{R}{\alpha}}(x_i) + (1-\mu_A(x_i))^{\frac{R}{\alpha}}\left(\frac{1}{2}\right)^{1-\frac{R}{\alpha}})^{\frac{\alpha}{R}} - 1\right] \\ &= \frac{R}{R-\alpha}\left(\frac{1}{2}\right)^{(1-\frac{R}{\alpha})\frac{\alpha}{R}}\sum_{i=1}^{n}\left[(\mu_A^{\frac{R}{\alpha}}(x_i)(x_i) + (1-\mu_A(x_i))^{\frac{R}{\alpha}})^{\frac{\alpha}{R}}\right] - \frac{nR}{R-\alpha} \end{aligned} \tag{2.54}$$

From Eq. 2.8,

$$H_R^\alpha(A) = \frac{R}{R-\alpha}\sum_{i=1}^{n}\left[1 - (\mu_A^{\alpha}{}^{\frac{R}{\alpha}}(x_i) + (1-\mu_A(x_i))^{\frac{R}{\alpha}})^{\frac{\alpha}{R}}\right]$$
$$\Rightarrow H_R^\alpha(A) = \frac{Rn}{R-\alpha} - \frac{R}{R-\alpha}\sum_{i=1}^{n}\left[(\mu_A^{\alpha}{}^{\frac{R}{\alpha}}(x_i) + (1-\mu_A(x_i))^{\frac{R}{\alpha}})^{\frac{\alpha}{R}}\right] \tag{2.55}$$

Using Eq. 2.55 in Eq. 2.54 we get

$$I_R^\alpha\left(A,\left[\frac{1}{2}\right]\right) = \left(\frac{1}{2}\right)^{\frac{\alpha-R}{R}}\left[\frac{Rn}{R-\alpha} - H_R^\alpha(A)\right] - \frac{nR}{R-\alpha}$$
$$\Rightarrow I_R^\alpha\left(A,\left[\frac{1}{2}\right]\right) = \left[\left(\frac{1}{2}\right)^{\frac{\alpha-R}{R}} - 1\right]\frac{nR}{R-\alpha} - \left(\frac{1}{2}\right)^{\frac{\alpha-R}{R}} H_R^\alpha(A).$$

Hence, 2.10 holds.

Theorem 2.11 *For fuzzy sets A and B, the relation between $H_R^m(A)$ and $I_R^m(A,B)$ is given by*

$$I_R^m\left(A,\left[\frac{1}{2}\right]\right) = \frac{1}{2}\left[\frac{-n(R-m+1)}{R-m} - H_R^m(A)\right].$$

Proof We know from Eq. 2.23 that

$$I_R^m(A,B) = \frac{R-m+1}{R-m}\sum_{i=1}^{n}\left[\begin{array}{l}\{\mu_A^{R-m+1}(x_i)\mu_B^{R-m+1}(x_i)\\ \quad + (1-\mu_A(x_i))^{R-m+1}(1-\mu_B(x_i))^{R-m+1}\}^{\frac{1}{R-m+1}} - 1\end{array}\right]$$

$$I_R^m\left(A,\left[\frac{1}{2}\right]\right) = \frac{R-m+1}{R-m}\sum_{i=1}^{n}\left[\begin{array}{l}\{\mu_A^{R-m+1}(x_i)\left(\frac{1}{2}\right)^{R-m+1}\\ \quad + (1-\mu_A(x_i))^{R-m+1}\left(\frac{1}{2}\right)^{R-m+1}\}^{\frac{1}{R-m+1}} - 1\end{array}\right]$$
$$= \frac{R-m+1}{R-m}\left(\frac{1}{2}\right)\sum_{i=1}^{n}\left[\begin{array}{l}\{\mu_A^{R-m+1}(x_i)\\ \quad + (1-\mu_A(x_i))^{R-m+1}\}^{\frac{1}{R-m+1}}\end{array}\right] - \frac{n(R-m+1)}{R-m} \tag{2.56}$$

From Eq. 2.9

$$
\begin{aligned}
H_R^m(A) &= \frac{R-m+1}{R-m}\sum_{i=1}^{n}\left[1-(\mu_A^{R-m+1}(x_i)+(1-\mu_A(x_i))^{R-m+1})^{\frac{1}{R-m+1}}\right] \\
&\Rightarrow H_R^m(A) \\
&= \frac{n(R-m+1)}{R-m}-\frac{R-m+1}{R-m}\sum_{i=1}^{n}\left[(\mu_A^{R-m+1}(x_i)+(1-\mu_A(x_i))^{R-m+1})^{\frac{1}{R-m+1}}\right]
\end{aligned}
\tag{2.57}
$$

Using Eq. 2.57 in Eq. 2.56 we get

$$
I_R^m\left(A,\left[\frac{1}{2}\right]\right)=\left(\frac{1}{2}\right)\left[\frac{n(R-m+1)}{R-m}-H_R^m(A)\right]-\frac{n(R-m+1)}{R-m}
$$

$$
\Rightarrow I_R^m\left(A,\left[\frac{1}{2}\right]\right)=\frac{1}{2}\left[\frac{-n(R-m+1)}{R-m}-H_R^m(A)\right].
$$

Hence, 2.11 holds.

2.8 Concluding Remarks

In this chapter, we have proposed two new parametric generalizations of one of the existing R-norm fuzzy information measures along with proof of their validity. It is noted that the proposed generalized fuzzy measures of information are valid measures which reduce to the known measure on substituting the particular values of parameters. Some of the interesting properties of these measures have also been studied. The similarity of proposed generalized measures of fuzzy information with $H_R(A)$ is proved using a numerical example. In addition, two new parametric generalized R-norm fuzzy divergence measures are introduced along with the proof of their validity. The interesting properties of these proposed generalized R-norm fuzzy divergence measures are also established. Finally, the relation between the proposed generalized R-norm fuzzy entropy and divergence measures are introduced. In view of application, the proposed measures are more flexible.

References

1. Arimoto SC (1971) Information theoretical considerations on estimation problems. Inf Control 19:181–194
2. Bhandari D, Pal NR (1993) Some new information measures for fuzzy sets. Inf Sci 67 (3):209–228
3. Boekee DE, Van der lubbe JCA (1980) The R-norm information measure. Inf Control 45:136–145

4. De Luca A, Termini S (1972) A definition of non-probabilistic entropy in the setting of fuzzy set theory. Inf Control 20(4):301–312
5. Hooda DS (2004) On generalized measures of fuzzy entropy. Mathematica Slovaca 54:315–325
6. Hooda DS, Bajaj RK (2008) On generalized *R*-norm information measures of fuzzy information. J Appl Math Stat Inform 4(2):199–212
7. Hooda DS, Jain D (2011) Generalized *R*-norm fuzzy information measures. J Appl Math Stat Inform 7(2):1–10
8. Hooda DS, Ram A (2002) Characterization of a generalized measure of *R*-norm entropy. Caribbean J Math Comput Sci 8:18–31
9. Hooda DS, Sharma DK (2008) Generalized *R*-norm information measures. J Appl Math Stat Inform 4(2):153–168
10. Kullback S, Leibler RA (1951) On information and sufficiency. Ann Math Stat 22(1):79–86
11. Kumar S (2009) Some more results on *R*-norm information measure. Tamkang J Math 40 (1):41–58
12. Kumar S, Choudhary A (2012) Generalized parametric *R*-norm information measure. Trends Appl Sci Res 7:350–369
13. Pal NR, Pal SK (1989) Object background segmentation using new definition of entropy. IEE Proc Comput Digital Tech 136(4):248–295
14. Shannon CE (1948) The mathematical theory of communication. Bell Syst Tech J 27(3):379–423
15. Zadeh LA (1968) Probability measures of fuzzy events. J Math Anal Appl 23:421–427

Chapter 3
Parametric Generalized Exponential Fuzzy Divergence Measure and Strategic Decision-Making

This chapter introduces and details a generalized methodology for measuring the degree of difference between two fuzzy sets. We present a new parametric generalized exponential measure of fuzzy divergence and study the essential properties of this measure in order to check its authenticity.

Section 3.1 gives a discussion on some well-known concepts and notations related to fuzzy set theory and fuzzy divergence measures. Thereafter, we introduce a parametric generalized fuzzy exponential measure of divergence corresponding to generalized fuzzy entropy given by Verma and Sharma [25] in Sect. 3.2. In Sect. 3.3, we first provide some interesting properties of the proposed measure of fuzzy divergence and then a relation is established between generalized exponential fuzzy entropy and the proposed fuzzy divergence measure. In Sect. 3.4, a comparison of the proposed divergence with some of existing generalized measures of fuzzy divergence is presented with the help of table and graph. The application of the proposed parametric generalized exponential measure of fuzzy divergence to strategic decision-making is illustrated with the help of a numerical example in Sect. 3.5. Section 3.6 presents the application of the proposed measure of fuzzy divergence in the existing methods of strategic decision-making. In the same section, a comparative analysis between the proposed method of strategic decision-making and the existing methods is provided. The final section concludes the chapter.

3.1 Generalized Measures of Fuzzy Divergence

This section is devoted to review some well-known notations and to recall the axiomatic definition of a divergence measure for fuzzy sets.

As we have mentioned in Sect. 1.1.4 of Chap. 1 that Zadeh [26] gave some notions related to fuzzy sets. Some of them, which we shall need in our discussion, are as follows:

A. Ohlan and R. Ohlan, *Generalizations of Fuzzy Information Measures*,
DOI 10.1007/978-3-319-45928-8_3

(1) **Complement**: $\overline{A}$ = Compliment of $A \Leftrightarrow \mu_{\overline{A}}(x) = 1 - \mu_A(x)$ for all $x \in X$.
(2) **Union:** $A \cup B$ = Union of A and $B \Leftrightarrow \mu_{A \cup B}(x) = \max\{\mu_A(x), \mu_B(x)\}$ for all $x \in X$.
(3) **Intersection**: $A \cap B$ = Intersection of A and $B \Leftrightarrow \mu_{A \cap B}(x) = \min\{\mu_A(x), \mu_B(x)\}$ for all $x \in X$.

Fuzzy Divergence Measures

In fuzzy context, several measures have been proposed in order to measure the degree of difference between two fuzzy sets. A general study of the axiomatic definition of a divergence measure for fuzzy sets was presented in Bouchon-Meunier et al. [5] and as a particular case it was studied widely in Montes et al. [19].

Bhandari and Pal [2] introduced the measure of fuzzy divergence corresponding to Kullback and Leibler [18] measure of divergence, as

$$I(A : B) = \sum_{i=1}^{n} \left[\mu_A(x_i) \log \frac{\mu_A(x_i)}{\mu_B(x_i)} + (1 - \mu_A(x_i)) \log \frac{1 - \mu_A(x_i)}{1 - \mu_B(x_i)} \right] \tag{3.1}$$

and also provided the essential conditions for a measure of divergence.

Measure of fuzzy divergence between two fuzzy sets gives the difference between two fuzzy sets and this measure of difference between two fuzzy sets is called the fuzzy divergence measure.

Fan and Xie [9] proposed the fuzzy information of discrimination of a fuzzy set A against other fuzzy set B corresponding to the exponential fuzzy entropy of Pal and Pal [20] and is defined by

$$I(A, B) = \sum_{i=1}^{n} \left[1 - (1 - \mu_A(x_i)) e^{\mu_A(x_i) - \mu_B(x_i)} - \mu_A(x_i) e^{(\mu_B(x_i) - \mu_A(x_i))} \right] \tag{3.2}$$

Finally, we mention some other generalized measures of fuzzy divergence with which we compare our measure.

Kapur [17] presented a fuzzy divergence measure corresponding to Havrada and Charvat [14] measure of divergence given by

$$\overline{I_\alpha}(A, B) = \frac{1}{\alpha - 1} \sum_{i=1}^{n} \left[\mu_A^{\alpha}(x_i) \mu_B^{1-\alpha}(x_i) + (1 - \mu_A(x_i))^{\alpha} (1 - \mu_B(x_i))^{1-\alpha} - 1 \right], \quad \alpha \neq 1, \alpha > 0 \tag{3.3}$$

Parkash et al. [21] proposed a fuzzy divergence measure corresponding to Ferreri's [11] probabilistic measure of divergence given by

$$I_a(A:B) = \sum_{i=1}^{n}\left[\mu_A(x_i)\log\frac{\mu_A(x_i)}{\mu_B(x_i)} + (1-\mu_A(x_i))\log\frac{1-\mu_A(x_i)}{1-\mu_B(x_i)}\right]$$
$$-\frac{1}{a}\sum_{i=1}^{n}\left[(1+a\mu_A(x_i))\log\frac{1+a\mu_A(x_i)}{1+a\mu_B(x_i)} + \{1+a(1-\mu_A(x_i))\}\log\frac{1+a(1-\mu_A(x_i))}{1+a(1-\mu_B(x_i))}\right] \tag{3.4}$$

Corresponding to Renyi [23] generalized measure of divergence Bajaj and Hooda [1] provided the generalized fuzzy divergence measure given by

$$D_\alpha(A,B) = \frac{1}{\alpha-1}\sum_{i=1}^{n}\log\left[\mu_A^\alpha(x_i)\mu_B^{1-\alpha}(x_i) + (1-\mu_A(x_i))^\alpha(1-\mu_B(x_i))^{1-\alpha}\right], \quad \alpha \neq 1,\ \alpha > 0. \tag{3.5}$$

3.2 New Parametric Generalized Exponential Measure of Fuzzy Divergence

Let us now turn to propose a new parametric generalized exponential measure of divergence between fuzzy sets A and B of universe of discourse $X = \{x_1, x_2, \ldots, x_n\}$ having the membership values $\mu_A(x_i), \mu_B(x_i), i = 1, 2, \ldots, n$, corresponding to generalized exponential fuzzy entropy of order $\alpha > 0$ given by Verma and Sharma [25] as

$$I_{E_\alpha}(A,B) = \sum_{i=1}^{n}\left[1 - (1-\mu_A(x_i))e^{((1-\mu_B(x_i))^\alpha - (1-\mu_A(x_i))^\alpha)} - \mu_A(x_i)e^{(\mu_B^\alpha(x_i) - \mu_A^\alpha(x_i))}\right] \tag{3.6}$$

Theorem 3.1 *$I_{E_\alpha}(A,B)$ is a valid measure of fuzzy divergence.*

Proof It is clear from (3.6) that
(i) $I_{E_\alpha}(A,B) \geq 0$
(ii) $I_{E_\alpha}(A,B) = 0$ if $\mu_A(x_i) = \mu_B(x_i)$, $\forall i = 1, 2, \ldots, n$
(iii) We now check the convexity of $I_{E_\alpha}(A,B)$:

$$\frac{\partial I_{E_\alpha}}{\partial \mu_A(x_i)} = (1-\alpha(1-\mu_A(x_i))^\alpha)e^{((1-\mu_B(x_i))^\alpha - (1-\mu_A(x_i))^\alpha)} + (\alpha-1)e^{(\mu_B^\alpha(x_i) - \mu_A^\alpha(x_i))}$$

$$\frac{\partial^2 I_{E_\alpha}}{\partial \mu_A^2(x_i)} = \alpha(\alpha+1)\left[(1-\mu_A(x_i))^{\alpha-1}e^{(1-\mu_B(x_i))^\alpha - (1-\mu_A(x_i))^\alpha} + \mu_A^{\alpha-1}(x_i)e^{\mu_B^\alpha(x_i) - \mu_A^\alpha(x_i)}\right]$$
$$-\alpha^2\left[(1-\mu_A(x_i))^{2\alpha-1}e^{(1-\mu_B(x_i))^\alpha - (1-\mu_A(x_i))^\alpha} + \mu_A^{2\alpha-1}(x_i)e^{\mu_B^\alpha(x_i) - \mu_A^\alpha(x_i)}\right] > 0$$

for $\alpha > 0$.

Similarly, $\frac{\partial^2 I_{E_\alpha}}{\partial \mu_B^2(x_i)} > 0$ for $\alpha > 0$.

Thus $I_{E_\alpha}(A,B)$ is a convex function of fuzzy sets A and B and hence in view of the definition of fuzzy divergence measure of Bhandari and Pal [2] provided in Sect. 1.4 of Chap. 1, $I_{E_\alpha}(A,B)$ is a valid measure of fuzzy divergence and consequently it establishes the validity of $J_{E_\alpha}(A,B) = I_{E_\alpha}(A,B) + I_{E_\alpha}(B,A)$ as a new parametric generalized exponential symmetric measure of divergence between fuzzy sets A and B.

In particular,

For $\alpha = 1$, $I_{E_\alpha}(A,B)$ reduces to $I(A,B)$ given in (3.2).

3.3 Properties of Generalized Exponential Fuzzy Divergence Measure

The generalized exponential fuzzy divergence measure $I_{E_\alpha}(A,B)$ defined above has the following properties. While proving these theorems we consider the separation of X into two parts X_1 and X_2, such that the set

$$X_1 = \{x/x \in X, \mu_A(x_i) \geq \mu_B(x_i)\}$$

and

$$X_2 = \{x/x \in X, \mu_A(x_i) < \mu_B(x_i)\}.$$

Using the notions explained in Sect. 3.1, we get

In set X_1,

$$A \cup B = \text{Union of } A \text{ and } B \Leftrightarrow \mu_{A\cup B}(x) = \max\{\mu_A(x), \mu_B(x)\} = \mu_A(x)$$
$$A \cap B = \text{Intersection of } A \text{ and } B \Leftrightarrow \mu_{A\cap B}(x) = \min\{\mu_A(x), \mu_B(x)\} = \mu_B(x).$$

In set X_2,

$$A \cup B = \text{Union of } A \text{ and } B \Leftrightarrow \mu_{A\cup B}(x) = \max\{\mu_A(x), \mu_B(x)\} = \mu_B(x)$$
$$A \cap B = \text{Intersection of } A \text{ and } B \Leftrightarrow \mu_{A\cap B}(x) = \min\{\mu_A(x), \mu_B(x)\} = \mu_A(x).$$

Theorem 3.2

(a) $I_{E_\alpha}(A\cup B, A) + I_{E_\alpha}(A\cap B, A) = I_{E_\alpha}(B, A)$.

(b) $I_{E_\alpha}(A\cup B, C) + I_{E_\alpha}(A\cap B, C) = I_{E_\alpha}(A, C) + I_{E_\alpha}(B, C)$.

(c) $I_{E_\alpha}(\overline{A\cup B}, \overline{A\cap B}) = I_{E_\alpha}(\overline{A}\cap\overline{B}, \overline{A}\cup\overline{B})$.

Proof 3.2(a)

$$
\begin{aligned}
I_{E_\alpha}(A \cup B, A) &= \sum_{i=1}^{n}\left[1-(1-\mu_{A\cup B}(x_i))e^{((1-\mu_A(x_i))^\alpha-(1-\mu_{A\cup B}(x_i))^\alpha)}-\mu_{A\cup B}(x_i)e^{(\mu_A^\alpha(x_i)-\mu_{A\cup B}^\alpha(x_i))}\right]\\
&= \sum_{X_1}\left[1-(1-\mu_A(x_i))e^{((1-\mu_A(x_i))^\alpha-(1-\mu_A(x_i))^\alpha)}-\mu_A(x_i)e^{(\mu_A^\alpha(x_i)-\mu_A^\alpha(x_i))}\right]\\
&\quad+\sum_{X_2}\left[1-(1-\mu_B(x_i))e^{((1-\mu_A(x_i))^\alpha-(1-\mu_B(x_i))^\alpha)}-\mu_B(x_i)e^{(\mu_A^\alpha(x_i)-\mu_B^\alpha(x_i))}\right]\\
&= \sum_{X_2}\left[1-(1-\mu_B(x_i))e^{((1-\mu_A(x_i))^\alpha-(1-\mu_B(x_i))^\alpha)}-\mu_B(x_i)e^{(\mu_A^\alpha(x_i)-\mu_B^\alpha(x_i))}\right]
\end{aligned}
\tag{3.7}
$$

Now

$$
\begin{aligned}
I_{E_\alpha}(A \cap B, A) &= \sum_{i=1}^{n}\left[1-(1-\mu_{A\cap B}(x_i))e^{((1-\mu_A(x_i))^\alpha-(1-\mu_{A\cap B}(x_i))^\alpha)}-\mu_{A\cap B}(x_i)e^{(\mu_A^\alpha(x_i)-\mu_{A\cap B}^\alpha(x_i))}\right]\\
&= \sum_{X_1}\left[1-(1-\mu_B(x_i))e^{((1-\mu_A(x_i))^\alpha-(1-\mu_B(x_i))^\alpha)}-\mu_B(x_i)e^{(\mu_A^\alpha(x_i)-\mu_B^\alpha(x_i))}\right]\\
&\quad+\sum_{X_2}\left[1-(1-\mu_A(x_i))e^{((1-\mu_A(x_i))^\alpha-(1-\mu_A(x_i))^\alpha)}-\mu_A(x_i)e^{(\mu_A^\alpha(x_i)-\mu_A^\alpha(x_i))}\right]\\
&= \sum_{X_1}\left[1-(1-\mu_B(x_i))e^{((1-\mu_A(x_i))^\alpha-(1-\mu_B(x_i))^\alpha)}-\mu_B(x_i)e^{(\mu_A^\alpha(x_i)-\mu_B^\alpha(x_i))}\right]
\end{aligned}
\tag{3.8}
$$

Adding Eqs. 3.7 and 3.8 we get

$$
\begin{aligned}
I_{E_\alpha}(A \cup B, A) + I_{E_\alpha}(A \cap B, A) &= \sum_{X_1}\left[1-(1-\mu_B(x_i))e^{((1-\mu_A(x_i))^\alpha-(1-\mu_B(x_i))^\alpha)}-\mu_B(x_i)e^{(\mu_A^\alpha(x_i)-\mu_B^\alpha(x_i))}\right]\\
&\quad+\sum_{X_2}\left[1-(1-\mu_B(x_i))e^{((1-\mu_A(x_i))^\alpha-(1-\mu_B(x_i))^\alpha)}-\mu_B(x_i)e^{(\mu_A^\alpha(x_i)-\mu_B^\alpha(x_i))}\right]\\
&= \sum_{i=1}^{n}\left[1-(1-\mu_B(x_i))e^{((1-\mu_A(x_i))^\alpha-(1-\mu_B(x_i))^\alpha)}-\mu_B(x_i)e^{(\mu_A^\alpha(x_i)-\mu_B^\alpha(x_i))}\right]\\
&= I_{E_\alpha}(B, A)
\end{aligned}
$$

Thus, $I_{E_\alpha}(A \cup B, A) + I_{E_\alpha}(A \cap B, A) = I_{E_\alpha}(B, A)$.
Hence, 3.2(a) holds.

3.2(b) $I_{E_\alpha}(A \cup B, C)$

$$= \sum_{i=1}^{n} \left[1 - (1 - \mu_{A\cup B}(x_i)) e^{((1-\mu_C(x_i))^\alpha - (1-\mu_{A\cup B}(x_i))^\alpha)} - \mu_{A\cup B}(x_i) e^{(\mu_C^\alpha(x_i) - \mu_{A\cup B}^\alpha(x_i))} \right]$$
$$= \sum_{X_1} \left[1 - (1 - \mu_A(x_i)) e^{((1-\mu_C(x_i))^\alpha - (1-\mu_A(x_i))^\alpha)} - \mu_A(x_i) e^{(\mu_C^\alpha(x_i) - \mu_A^\alpha(x_i))} \right]$$
$$+ \sum_{X_2} \left[1 - (1 - \mu_B(x_i)) e^{((1-\mu_C(x_i))^\alpha - (1-\mu_B(x_i))^\alpha)} - \mu_B(x_i) e^{(\mu_C^\alpha(x_i) - \mu_B^\alpha(x_i))} \right] \tag{3.9}$$

$I_{E_\alpha}(A \cap B, C)$

$$= \sum_{i=1}^{n} \left[1 - (1 - \mu_{A\cap B}(x_i)) e^{((1-\mu_C(x_i))^\alpha - (1-\mu_{A\cap B}(x_i))^\alpha)} - \mu_{A\cap B}(x_i) e^{(\mu_C^\alpha(x_i) - \mu_{A\cap B}^\alpha(x_i))} \right]$$
$$= \sum_{X_1} \left[1 - (1 - \mu_B(x_i)) e^{((1-\mu_C(x_i))^\alpha - (1-\mu_B(x_i))^\alpha)} - \mu_B(x_i) e^{(\mu_C^\alpha(x_i) - \mu_B^\alpha(x_i))} \right]$$
$$+ \sum_{X_2} \left[1 - (1 - \mu_A(x_i)) e^{((1-\mu_C(x_i))^\alpha - (1-\mu_A(x_i))^\alpha)} - \mu_A(x_i) e^{(\mu_C^\alpha(x_i) - \mu_A^\alpha(x_i))} \right] \tag{3.10}$$

Adding Eqs. 3.9 and 3.10 we get

$$I_{E_\alpha}(A \cup B, C) + I_{E_\alpha}(A \cap B, C)$$
$$= \sum_{i=1}^{n} \left[1 - (1 - \mu_A(x_i)) e^{((1-\mu_C(x_i))^\alpha - (1-\mu_A(x_i))^\alpha)} - \mu_A(x_i) e^{(\mu_C^\alpha(x_i) - \mu_A^\alpha(x_i))} \right]$$
$$+ \sum_{i=1}^{n} \left[1 - (1 - \mu_B(x_i)) e^{((1-\mu_C(x_i))^\alpha - (1-\mu_B(x_i))^\alpha)} - \mu_B(x_i) e^{(\mu_C^\alpha(x_i) - \mu_B^\alpha(x_i))} \right]$$
$$= I_{E_\alpha}(A, C) + I_{E_\alpha}(B, C).$$

Hence, 3.2(b) holds.

3.2(c) $I_{E_\alpha}(\overline{A \cup B}, \overline{A \cap B})$

$$= \sum_{i=1}^{n} \left[1 - (1 - \mu_{\overline{A\cup B}}(x_i)) e^{((1-\mu_{\overline{A\cap B}}(x_i))^\alpha - (1-\mu_{\overline{A\cup B}}(x_i))^\alpha)} - \mu_{\overline{A\cup B}}(x_i) e^{(\mu_{\overline{A\cap B}}^\alpha(x_i) - \mu_{\overline{A\cup B}}^\alpha(x_i))} \right]$$
$$= \sum_{i=1}^{n} \left[1 - \mu_{A\cup B}(x_i) e^{(\mu_{A\cap B}^\alpha(x_i) - \mu_{A\cup B}^\alpha(x_i))} - (1 - \mu_{A\cup B}(x_i)) e^{((1-\mu_{A\cap B}(x_i))^\alpha - (1-\mu_{A\cup B}(x_i))^\alpha)} \right]$$
$$= \sum_{X_1} \left[1 - \mu_A(x_i) e^{(\mu_B^\alpha(x_i) - \mu_A^\alpha(x_i))} - (1 - \mu_A(x_i)) e^{((1-\mu_B(x_i))^\alpha - (1-\mu_A(x_i))^\alpha)} \right]$$
$$+ \sum_{X_2} \left[1 - \mu_B(x_i) e^{(\mu_A^\alpha(x_i) - \mu_B^\alpha(x_i))} - (1 - \mu_B(x_i)) e^{((1-\mu_A(x_i))^\alpha - (1-\mu_B(x_i))^\alpha)} \right]$$

and

$$I_{E_\alpha}(\overline{A}\cap\overline{B},\overline{A}\cup\overline{B})$$
$$=\sum_{i=1}^{n}\left[1-(1-\mu_{\overline{A}\cap\overline{B}}(x_i))e^{((1-\mu_{\overline{A}\cup\overline{B}}(x_i))^\alpha-(1-\mu_{\overline{A}\cap\overline{B}}(x_i))^\alpha)}-\mu_{\overline{A}\cap\overline{B}}(x_i)e^{(\mu^\alpha_{\overline{A}\cup\overline{B}}(x_i)-\mu^\alpha_{\overline{A}\cap\overline{B}}(x_i))}\right]$$
$$=\sum_{X_1}\left[1-(1-\mu_{\overline{A}}(x_i))e^{((1-\mu_{\overline{B}}(x_i))^\alpha-(1-\mu_{\overline{A}}(x_i))^\alpha)}-\mu_{\overline{A}}(x_i)e^{(\mu^\alpha_{\overline{B}}(x_i)-\mu^\alpha_{\overline{A}}(x_i))}\right]$$
$$+\sum_{X_2}\left[1-(1-\mu_{\overline{B}}(x_i))e^{((1-\mu_{\overline{A}}(x_i))^\alpha-(1-\mu_{\overline{B}}(x_i))^\alpha)}-\mu_{\overline{B}}(x_i)e^{(\mu^\alpha_{\overline{A}}(x_i)-\mu^\alpha_{\overline{B}}(x_i))}\right]$$
$$=\sum_{X_1}\left[1-\mu_A(x_i)e^{(\mu^\alpha_B(x_i)-\mu^\alpha_A(x_i))}-(1-\mu_A(x_i))e^{((1-\mu_B(x_i))^\alpha-(1-\mu_A(x_i))^\alpha)}\right]$$
$$+\sum_{X_2}\left[1-\mu_B(x_i)e^{(\mu^\alpha_A(x_i)-\mu^\alpha_B(x_i))}-(1-\mu_B(x_i))e^{((1-\mu_A(x_i))^\alpha-(1-\mu_B(x_i))^\alpha)}\right]$$
$$=I_{E_\alpha}(\overline{A\cup B},\overline{A\cap B})$$

Thus, $I_{E_\alpha}(\overline{A\cup B},\overline{A\cap B})=I_{E_\alpha}(\overline{A}\cap\overline{B},\overline{A}\cup\overline{B})$.
Hence, 3.2(c) holds.

Theorem 3.3
(a) $I_{E_\alpha}(A,\overline{A})=I_{E_\alpha}(\overline{A},A)$.
(b) $I_{E_\alpha}(\overline{A},\overline{B})=I_{E_\alpha}(A,B)$.
(c) $I_{E_\alpha}(A,\overline{B})=I_{E_\alpha}(\overline{A},B)$.
(d) $I_{E_\alpha}(A,B)+I_{E_\alpha}(\overline{A},B)=I_{E_\alpha}(\overline{A},\overline{B})+I_{E_\alpha}(A,\overline{B})$.

Proof

3.3(a) $I_{E_\alpha}(A,\overline{A})$
$$=\sum_{i=1}^{n}\left[1-(1-\mu_A(x_i))e^{((1-\mu_{\overline{A}}(x_i))^\alpha-(1-\mu_A(x_i))^\alpha)}-\mu_A(x_i)e^{(\mu^\alpha_{\overline{A}}(x_i)-\mu^\alpha_A(x_i))}\right]$$
$$=\sum_{i=1}^{n}\left[1-(1-\mu_A(x_i))e^{\mu^\alpha_A(x_i)-(1-\mu_A(x_i))^\alpha)}-\mu_A(x_i)e^{((1-\mu_A(x_i))^\alpha-\mu^\alpha_A(x_i))}\right]$$

and

$$I_{E_\alpha}(\overline{A},A)$$
$$=\sum_{i=1}^{n}\left[1-(1-\mu_{\overline{A}}(x_i))e^{((1-\mu_A(x_i))^\alpha-(1-\mu_{\overline{A}}(x_i))^\alpha)}-\mu_{\overline{A}}(x_i)e^{(\mu^\alpha_A(x_i)-\mu^\alpha_{\overline{A}}(x_i))}\right]$$
$$=\sum_{i=1}^{n}\left[1-\mu_A(x_i)e^{((1-\mu_A(x_i))^\alpha-\mu^\alpha_A(x_i))}-(1-\mu_A(x_i))e^{\mu^\alpha_A(x_i)-(1-\mu_A(x_i))^\alpha)}\right]$$
$$=I_{E_\alpha}(A,\overline{A})$$

Thus, $I_{E_\alpha}(A,\overline{A}) = I_{E_\alpha}(\overline{A},A)$.
Hence, 3.3(a) holds.

3.3(b) $I_{E_\alpha}(\overline{A},\overline{B})$

$$= \sum_{i=1}^{n}\left[1-(1-\mu_{\overline{A}}(x_i))e^{((1-\mu_{\overline{B}}(x_i))^\alpha-(1-\mu_{\overline{A}}(x_i))^\alpha)} - \mu_{\overline{A}}(x_i)e^{(\mu^\alpha_{\overline{B}}(x_i)-\mu^\alpha_{\overline{A}}(x_i))}\right]$$

$$= \sum_{i=1}^{n}\left[1-\mu_A(x_i)e^{(\mu^\alpha_B(x_i)-\mu^\alpha_A(x_i))} - (1-\mu_A(x_i))e^{((1-\mu_B(x_i))^\alpha-(1-\mu_A(x_i))^\alpha)}\right]$$

$$= I_{E_\alpha}(A,B)$$

Thus, $I_{E_\alpha}(\overline{A},\overline{B}) = I_{E_\alpha}(A,B)$.
Hence, 3.3(b) holds.

3.3(c) $I_{E_\alpha}(A,\overline{B})$

$$= \sum_{i=1}^{n}\left[1-(1-\mu_A(x_i))e^{((1-\mu_{\overline{B}}(x_i))^\alpha-(1-\mu_A(x_i))^\alpha)} - \mu_A(x_i)e^{(\mu^\alpha_{\overline{B}}(x_i)-\mu^\alpha_A(x_i))}\right]$$

$$= \sum_{i=1}^{n}\left[1-(1-\mu_A(x_i))e^{\mu^\alpha_B(x_i)-(1-\mu_A(x_i))^\alpha)} - \mu_A(x_i)e^{((1-\mu_B(x_i))^\alpha-\mu^\alpha_A(x_i))}\right]$$

Now $I_{E_\alpha}(\overline{A},B)$

$$= \sum_{i=1}^{n}\left[1-(1-\mu_{\overline{A}}(x_i))e^{((1-\mu_B(x_i))^\alpha-(1-\mu_{\overline{A}}(x_i))^\alpha)} - \mu_{\overline{A}}(x_i)e^{(\mu^\alpha_B(x_i)-\mu^\alpha_{\overline{A}}(x_i))}\right]$$

$$= \sum_{i=1}^{n}\left[1-\mu_A(x_i)e^{((1-\mu_B(x_i))^\alpha-\mu^\alpha_A(x_i))} - (1-\mu_A(x_i))e^{\mu^\alpha_B(x_i)-(1-\mu_A(x_i))^\alpha)}\right]$$

$$= I_{E_\alpha}(A,\overline{B})$$

Thus, $I_{E_\alpha}(A,\overline{B}) = I_{E_\alpha}(\overline{A},B)$.
Hence, 3.3(c) holds.
3.3(d) It obviously follows from 3.3(b) and 3.3(c).

Theorem 3.4 *Relation between $E_\alpha(A)$ and $I_{E_\alpha}(A,B)$ is given by*

$$E_\alpha(A) = 1 - \frac{e^{1-0.5^\alpha}}{n(e^{1-0.5^\alpha}-1)} I_{E_\alpha}\left(A, \left[\frac{1}{2}\right]\right).$$

Proof

$$
\begin{aligned}
I_{E_\alpha}\left(A, \left[\frac{1}{2}\right]\right) &= \sum_{i=1}^{n}\left[1-(1-\mu_A(x_i))e^{(0.5^\alpha-(1-\mu_A(x_i))^\alpha)} - \mu_A(x_i)e^{(0.5^\alpha-\mu_A^\alpha(x_i))}\right] \\
&= \sum_{i=1}^{n}\left[1-(1-\mu_A(x_i))\frac{e^{-(1-\mu_A(x_i))^\alpha}}{e^{-0.5^\alpha}} - \mu_A(x_i)\frac{e^{-\mu_A^\alpha(x_i)}}{e^{-0.5^\alpha}}\right] \\
&= \sum_{i=1}^{n}\left[1-(1-\mu_A(x_i))\frac{e^{1-(1-\mu_A(x_i))^\alpha}}{e^{1-0.5^\alpha}} - \mu_A(x_i)\frac{e^{1-\mu_A^\alpha(x_i)}}{e^{1-0.5^\alpha}}\right] \\
&= n - \frac{1}{e^{(1-0.5^\alpha)}}\sum_{i=1}^{n}\left[(1-\mu_A(x_i))e^{(1-(1-\mu_A(x_i))^\alpha)} + \mu_A(x_i)e^{(1-\mu_A^\alpha(x_i))} + 1 - 1\right] \\
&= n - \frac{1}{e^{(1-0.5^\alpha)}}\sum_{i=1}^{n}\left[\mu_A(x_i)e^{(1-\mu_A^\alpha(x_i))} + (1-\mu_A(x_i))e^{1-(1-\mu_A(x_i))^\alpha} - 1\right] - \frac{n}{e^{(1-0.5^\alpha)}} \\
&= n - \frac{1}{e^{(1-0.5^\alpha)}}n\left(e^{(1-0.5^\alpha)}-1\right)E_\alpha(A) - \frac{n}{e^{(1-0.5^\alpha)}} \\
&= n\left(1-\frac{1}{e^{(1-0.5^\alpha)}}\right) - \frac{1}{e^{(1-0.5^\alpha)}}n\left(e^{(1-0.5^\alpha)}-1\right)E_\alpha(A) \\
&= \frac{n\left(e^{(1-0.5^\alpha)}-1\right)}{e^{(1-0.5^\alpha)}} - \frac{n\left(e^{(1-0.5^\alpha)}-1\right)}{e^{(1-0.5^\alpha)}}E_\alpha(A) \\
&= (1-E_\alpha(A))\frac{n\left(e^{(1-0.5^\alpha)}-1\right)}{e^{(1-0.5^\alpha)}} \\
&\Rightarrow \frac{e^{(1-0.5^\alpha)}}{n(e^{(1-0.5^\alpha)}-1)}I_{E_\alpha}\left(A, \left[\frac{1}{2}\right]\right) \\
&= (1-E_\alpha(A))
\end{aligned}
$$

Thus, $E_\alpha(A) = 1 - \frac{e^{(1-0.5^\alpha)}}{n(e^{(1-0.5^\alpha)}-1)}I_{E_\alpha}\left(A, \left[\frac{1}{2}\right]\right)$.

Hence, 3.4 holds.

3.4 A Comparative Study

In this section, we demonstrate the efficiency of proposed fuzzy divergence by comparing it with some of existing fuzzy divergence measures. To do so, we present the comparative study of the proposed divergence measure with the existing fuzzy divergence measures given by Kapur [17], Parkash et al. [21] and Bajaj and Hooda [1] provided in Sect. 3.1.

Let A and B be any two fuzzy sets given as $A = \{0.1, 0.9, 0.5\}$, $B = \{0.6, 0.7, 0.1\}$.

The computed values of fuzzy divergence measures $\overline{I_\alpha}(A:B)$, $I_\alpha(A:B)$, $D_\alpha(A,B)$, $I_{E_\alpha}(A,B)$ are presented in Table 3.1.

Table 3.1 and Fig. 3.1 depict the minimization of degree of divergence of the proposed fuzzy measure. It is clear that the proposed fuzzy divergence measure is efficient than the existing fuzzy divergence measures.

Table 3.1 Computed values of fuzzy divergence measures: $\overline{I_\alpha}(A:B)$, $D_\alpha(A,B)$, $I_a(A:B)$, $I_{E_\alpha}(A,B)$

Fuzzy divergence measure	$a, \alpha = 0.1$	$a, \alpha = 0.4$	$a, \alpha = 0.5$	$a, \alpha = 0.6$	$a, \alpha = 0.9$
$\overline{I_\alpha}(A:B)$	0.1243	0.4743	0.5875	0.7006	1.4809
$D_\alpha(A,B)$	0.1276	0.5058	0.6273	0.7456	1.0766
$I_a(A:B)$	1.1734	1.0272	0.9966	0.9681	0.8933
$I_{E_\alpha}(A,B)$	0.0977	0.2406	0.2805	0.3017	0.3416

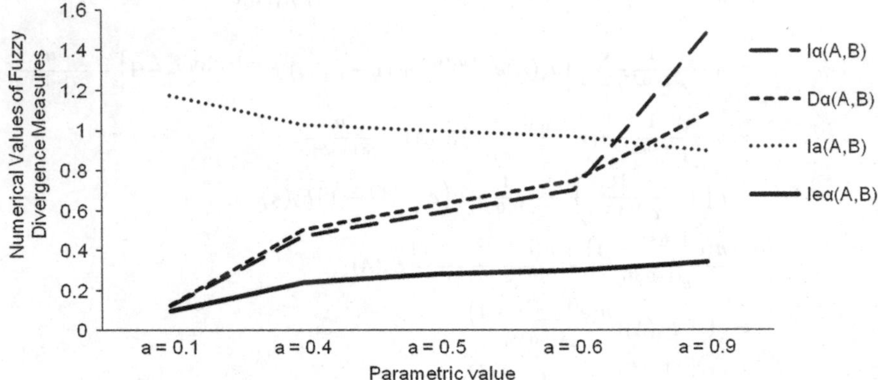

Fig. 3.1 Comparison of numerical values of fuzzy divergence measures: $\overline{I_\alpha}(A:B)$, $D_\alpha(A,B)$, $I_a(A:B)$, $I_{E_\alpha}(A,B)$

3.5 Application of Parametric Generalized Exponential Measure of Fuzzy Divergence in Strategic Decision-Making

As we have already discussed in Sect. 3.1 that in recent years the applications of the fuzzy divergence measure have been given in different areas: Poletti et al. [22] in bioinformatics; Bhandari et al. [3], Fan et al. [10] and Bhatia and Singh [4] in image thresholding; Ghosh et al. [13] in automated leukocyte recognition. We provide an application of the proposed divergence measure in strategic decision-making.

Decision-making problem is the process of finding the best option from all of the feasible alternatives. It is assumed that a firm X desires to apply m strategies $S_1, S_2, S_3, \ldots, S_m$ to meet its goal. Let each strategy has different degrees of effectiveness if the input associated with it is varied, let $\{I_1, I_2, I_3, \ldots, I_n\}$ be the input set or set of alternatives. Let the fuzzy set Y denotes the effectiveness of a particular strategy with uniform input. Therefore $Y = \{(Y, \mu_Y(S_i))/i = 1, 2, \ldots, m\}$. Further, let I_j be a fuzzy set denotes the degree of effectiveness of a strategy when it is implemented with input I_j.

$$I_j = \{(I_j, \mu_{I_j}(S_i)) \big/ i = 1, 2, \ldots, m\}$$

where $j = 1, 2, \ldots, n$.

Taking $A = Y$ and $B = I_j$ in the fuzzy divergence measure $I_{E_\alpha}(Y, I_j)$ defined in Sect. 3.3 we calculate $I_{E_\alpha}(Y, I_j)$. Then most effective I_j is determined by $\text{Min}\{I_{E_\alpha}(Y, I_j)\}_{\substack{1 \le j \le n \\ 0 < \alpha \le 0.5}}$. It is assumed that $I_t (1 \le t \le n)$ determines the minimum value of $\{I_{E_\alpha}(Y, I_j)\}_{0 < \alpha \le 0.5}$. With this I_t finds $\text{Max.}\{\mu_{I_j}(S_i)\}_{1 \le i \le m}$, let it corresponds to $S_p, 1 \le p \le m$. Hence, if the strategy S_p is implemented with input budget of I_t the firm will meet its goal in the most input-effective manner.

An Illustrative Example

Let $m = n = 5$ in the above model. Table 3.2 shows the efficiency of different strategies at uniform inputs.

Table 3.3 illustrates the efficiency of different strategies at particular inputs.

The numerical values of divergence measure $\{I_{E_\alpha}(Y, I_j)\}_{\substack{1 \le j \le n \\ 0 < \alpha \le 0.5}}$ are presented in Table 3.4. The calculated numerical values of the proposed fuzzy divergence measure indicate that alternative input I_2 is more appropriate for values of $\alpha (0 < \alpha \le 0.5)$. An examination of the results presented in Tables 3.3 and 3.4 makes it clear that the strategy S_3 is most effective. Thus, a firm will achieve its goal most effectively if the strategy S_3 is applied with an input alternative I_2.

Table 3.2 Efficiency of different strategies at uniform inputs

$\mu_Y(S_1)$	$\mu_Y(S_2)$	$\mu_Y(S_3)$	$\mu_Y(S_4)$	$\mu_Y(S_5)$
0.5	0.3	0.6	0.4	0.7

Table 3.3 Efficiency of strategies at particular inputs

I_j	$\mu_{I_j}(S_1)$	$\mu_{I_j}(S_2)$	$\mu_{I_j}(S_3)$	$\mu_{I_j}(S_4)$	$\mu_{I_j}(S_5)$
I_1	0.3	0.4	0.5	0.8	0.6
I_2	0.6	0.3	0.9	0.2	0.4
I_3	0.7	0.5	0.8	0.9	0.6
I_4	0.8	0.7	0.6	0.5	0.3
I_5	0.5	0.8	0.4	0.7	0.5

Table 3.4 Numerical values of divergence measure $I_{E_\alpha}(Y, I_j)\}_{\substack{1 \le j \le n \\ 0 < \alpha \le 0.5}}$

I_j	$\alpha = 0.1$	$\alpha = 0.3$	$\alpha = 0.4$	$\alpha = 0.5$
I_1	0.0456	0.0831	0.0900	0.0916
I_2	**0.0449**	**0.0662**	**0.0568**	**0.0396**
I_3	0.0779	0.1308	0.1289	0.1164
I_4	0.0748	0.1500	0.1646	0.1700
I_5	0.0764	0.1568	0.1748	0.1843

3.6 Application of Fuzzy TOPSIS and MOORA Methods for Strategic Decision-Making: A Comparative Analysis

We now present an application of TOPSIS method developed by Hwang and Yoon [15] and MOORA method developed by Brauers and Zavadskas [6] in a fuzzy context for strategic decision-making using the proposed fuzzy divergence measure (3.6). A methodology for solving decision-making problem using TOPSIS in fuzzy environment was developed by Chen [7], Jahanshahloo et al. [16] and Emami et al. [8]. Further, Stanujkic et al. [24] and Gadakh et al. [12] extended the MOORA method for solving decision-making problem with interval data and optimization problems in welding.

3.6.1 Fuzzy TOPSIS Method

Let us assume that there exists a set $I = \{I_1, I_2, I_3, \ldots, I_m\}$ of m alternative inputs and a set of n attributes (strategies) given by $S = \{S_1, S_2, S_3, \ldots, S_n\}$.The decision-maker has to find the best alternative from the set I corresponding to set S of n attributes (strategies).

The computational procedure of solving the fuzzy strategic decision-making problem involve the following steps:

1. Construct a fuzzy decision matrix.

$$\begin{array}{ccccc} & S_1 & S_2 & \cdots & S_n \\ I_1 & x_{11} & x_{12} & \cdots & x_{1n} \\ I_2 & x_{21} & x_{22} & \cdots & x_{2n} \\ \vdots & \vdots & \vdots & & \vdots \\ I_m & x_{m1} & x_{m2} & \cdots & x_{mn} \end{array}$$
$$W = [w_1, w_2, \ldots, w_n]$$

2. Construct the normalized fuzzy decision matrix. The normalized value n_{ij} is calculated as

$$n_{ji} = x_{ji} \Bigg/ \sqrt{\sum_{j=1}^{m} x_{ji}^2}, \quad j = 1, \ldots, m, \ i = 1, \ldots, n. \tag{3.11}$$

3. Construct the weighted normalized fuzzy decision matrix, the weighted normalized value

$$v_{ji} = w_i n_{ji}, \quad j = 1, \ldots, m, \ i = 1, \ldots, n \tag{3.12}$$

where the weighted matrix for each strategy is as follows: $W = [1, 1, 1, 1, 1]$ and w_i is the weight of i-th attribute.

4. Determine the fuzzy positive ideal and fuzzy negative ideal solution A^+ and A^-, using the formulae

$$A^+ = \{v_1^+, v_2^+, \ldots, v_n^+\} = \left\{ \max_j v_{ij} \Big/ i \in S \right\} \quad \text{and} \tag{3.13}$$

$$A^- = \{v_1^-, v_2^-, \ldots, v_n^-\} = \left\{ \min_j v_{ij} \Big/ i \in S \right\} \quad \text{respectively} \tag{3.14}$$

where S is associated with the set of different strategies.

5. Calculate the separation of each alternative input from positive ideal solution and negative ideal solution, respectively, using the proposed measure (3.6).
6. Calculate the relative closeness of each alternative to positive ideal solution using the formula

$$R_j = I_{E_\alpha}^- \Big/ \left(I_{E_\alpha}^+ + I_{E_\alpha}^- \right), \quad j = 1, \ldots, m. \tag{3.15}$$

7. Rank the preference order of all alternatives according to the closeness coefficient.

Now the application of proposed measure $I_{E_\alpha}(A, B)$ with TOPSIS technique is demonstrated using the fuzzy decision matrix considered in Table 3.3.

Table 3.5 presents the normalized/weighted fuzzy decision matrix corresponding to the fuzzy decision matrix given in Table 3.3 using the formulae (3.11) and (3.12).

Table 3.6 shows the fuzzy positive and negative ideal solutions A^+ and A^- using formulae (3.13) and (3.14).

Table 3.5 Normalized/weighted fuzzy decision matrix

	$\mu_{I_j}(S_1)$	$\mu_{I_j}(S_2)$	$\mu_{I_j}(S_3)$	$\mu_{I_j}(S_4)$	$\mu_{I_j}(S_5)$
I_1	0.2218	0.3133	0.3356	0.5010	0.5432
I_2	0.4435	0.2350	0.6040	0.1252	0.3622
I_3	0.5174	0.3916	0.5369	0.5636	0.5432
I_4	0.5914	0.5483	0.4027	0.3131	0.2716
I_5	0.3696	0.6266	0.2685	0.4383	0.4527

Table 3.6 Fuzzy positive and negative ideal solutions

	$\mu_{I_j}(S_1)$	$\mu_{I_j}(S_2)$	$\mu_{I_j}(S_3)$	$\mu_{I_j}(S_4)$	$\mu_{I_j}(S_5)$
A^+	0.5914	0.6266	0.6040	0.5636	0.5432
A^-	0.2218	0.2350	0.2685	0.1252	0.2716

The calculated numerical values of divergence values of each alternative input from positive ideal solution and negative ideal solution using measure (3.6) are given in Table 3.7.

The best alternative is the one with the shortest distance to the fuzzy positive ideal solution and with the longest distance to the fuzzy negative ideal solution. The calculated values of relative closeness of each alternative to positive ideal solution using formula (3.15) and their corresponding ranks are shown in Table 3.8.

According to the closeness coefficient, the ranking of the preference order of these alternatives $I_j (j = 1, 2, 3, 4, 5)$:

For $\alpha = 0.1$, $I_3 > I_5 > I_4 > I_1 > I_2$.
For $\alpha = 0.5$, $I_3 > I_5 > I_4 > I_1 > I_2$.

Thus, we here find that variation in values of α brings about change in ranking, but leaves the best choice unchanged. So I_3 is the most preferable alternative.

3.6.2 *Fuzzy MOORA Method*

Fuzzy MOORA method for solving strategic decision-making problems is as follows. The computational procedure in Fuzzy MOORA method up to Step 3 is same as discussed in Fuzzy TOPSIS method above.
Step 4. Compute the overall rating S_i^+ and S_i^- for different attributes (strategies) of each alternative using Table 3.5 and formulae

Table 3.7 Distance of each alternative from positive and negative ideal solutions

	$I_{E_\alpha}^+$		$I_{E_\alpha}^-$	
	$\alpha = 0.1$	$\alpha = 0.5$	$\alpha = 0.1$	$\alpha = 0.5$
I_1	0.1960	0.6621	0.1638	0.5308
I_2	0.2367	0.8460	0.1268	0.4294
I_3	0.0693	0.2505	0.2927	0.9497
I_4	0.1522	0.5319	0.2020	0.6662
I_5	0.1167	0.3974	0.2105	0.6866

Table 3.8 Closeness coefficient and ranking

Alternatives (Inputs)	$\alpha = 0.1$	Rank	$\alpha = 0.5$	Rank
I_1	0.4553	4	0.4450	4
I_2	0.3488	5	0.3367	5
I_3	**0.8086**	**1**	**0.7913**	**1**
I_4	0.5703	3	0.5560	3
I_5	0.6433	2	0.6334	2

$$S_i^+ = \{v_1^+, v_2^+, \ldots, v_m^+\} = \left\{ \max_i v_{ij} \Big/ j \in I \right\} \text{ and} \tag{3.16}$$

$$S_i^- = \{v_1^-, v_2^-, \ldots, v_m^-\} = \left\{ \min_i v_{ij} \Big/ j \in I \right\} \quad \text{respectively,} \tag{3.17}$$

where $J = \{j = 1, 2, 3, \ldots, m/j$ is associated with input$\}$.

The calculated overall rating S_i^+ and S_i^- for different attributes of each alternative using formulae (3.16) and (3.17) are presented in Table 3.9 as follows:
Step 5. Calculate the overall performance index $I_{E_\alpha}(S_i^+, S_i^-)$ for each alternative using the formula (3.6) and the computed values given in Table 3.9.
Step 6. Ranking alternatives and/or selecting the most efficient one are based on the values of $I_{E_\alpha}(S_i^+, S_i^-)$.

The overall performance index $I_{E_\alpha}(S_i^+, S_i^-)$ for each alternative is calculated using formula (3.6) and values obtained in Table 3.9. Finally, the ranking results have been obtained using MOORA method and are presented in Table 3.10.

According to the obtained results, ranking order of alternatives $I_j (j = 1, 2, 3, 4, 5)$ for $\alpha = 0.1$ and $\alpha = 0.5$, we here find that variation in values of α brings about change in ranking, but leaves the best choice unchanged. So I_3 is the most preferable alternative.

Table 3.9 The overall rating S_i^+ and S_i^- of each alternative

	S_i^+	S_i^-
I_1	0.5432	0.2218
I_2	0.6040	0.1252
I_3	0.5636	0.3916
I_4	0.5914	0.2716
I_5	0.6266	0.2685

Table 3.10 The ranking results obtained using MOORA method

	$I_{E_\alpha}(S_i^+, S_i^-)$			
	$\alpha = 0.1$	Rank	$\alpha = 0.5$	Rank
I_1	0.0184	3	0.0222	2
I_2	0.0472	5	0.0667	5
I_3	**0.0049**	**1**	**0.0109**	**1**
I_4	0.0173	2	0.0322	3
I_5	0.0225	4	0.0468	4

3.6.3 A Comparative Analysis of the Proposed Method and the Existing Methods of Strategic Decision-Making

We now compare the proposed method of strategic decision-making with the existing methods of strategic decision-making using the proposed measure (3.6). From the proposed method in Sect. 3.6, it is clear that a firm will achieve its goal most effectively if the strategy S_4 is applied with an input alternative I_2. Thus, I_2 is best input alternative for a firm to achieve its goal. However, we above examine from the existing methods that I_3 is a preferable input alternative. It is also noticed that the proposed method is very short, simple and consistent method than the existing methods of strategic decision-making having a computational procedure involving a number of steps. Thus, the proposed method of strategic decision-making is better than the existing methods.

3.7 Concluding Remarks

In this chapter, we have proposed and validated the generalized exponential measure of fuzzy divergence. A relation is established between generalized exponential fuzzy entropy and the proposed fuzzy divergence measure. Particular case and some of the properties are proved on this divergence measure. The efficiency of the proposed divergence measure is also presented. In addition, application of the proposed divergence measure is discussed in strategic decision-making and a numerical example is given for illustration. The application of the proposed measure of fuzzy divergence is presented in two existing methods of strategic decision-making. A comparative analysis between the proposed method of strategic decision-making and the existing methods of strategic decision-making has also been provided. It is noted that the proposed measure of fuzzy divergence comprises greater consistency and flexibility in applications because of the presence of the parameter.

References

1. Bajaj RK, Hooda DS (2010) On some new generalized measures of fuzzy information. World Acad Sci Eng Technol 62:747–753
2. Bhandari D, Pal NR (1993) Some new information measures for fuzzy sets. Inf Sci 67 (3):209–228
3. Bhandari D, Pal NR, Majumder DD (1992) Fuzzy divergence, probability measure of fuzzy event and image thresholding. Inf Sci 13:857–867

4. Bhatia PK, Singh S (2013) A new measure of fuzzy directed divergence and its application in image segmentation. Int J Intell Syst Appl 4:81–89
5. Bouchon-Meunier B, Rifqi M, Bothorel S (1996) Towards general measures of comparison of objects. Fuzzy Sets Syst 84:143–153
6. Brauers WKM, Zavadskas EK (2006) The MOORA method and its application to privatization in transition economy. Control Cybern 35(2):443–468
7. Chen CT (2000) Extension of TOPSIS for group decision making under fuzzy environment. Fuzzy Sets Syst 114(1):1–9
8. Emami M, Nazari K, Fardmanesh H (2012) Application of fuzzy TOPSIS technique for strategic management decision. J Appl Basic Sci Res 2(1):685–689
9. Fan J, Xie W (1999) Distance measures and induced fuzzy entropy. Fuzzy Sets Syst 104 (2):305–314
10. Fan S, Yang S, He P, Nie H (2011) Infrared electric image thresholding using two dimensional fuzzy entropy. Energy Procedia 12:411–419
11. Ferreri C (1980) Hyperentropy and related heterogeneity divergence and information measures. Statistica 40(2):155–168
12. Gadakh VS, Shinde VB, Khemnar NS (2013) Optimization of welding process parameters using MOORA method. Int J Adv Manuf Technol 69:2031–2039
13. Ghosh M, Das D, Chakraborty C, Ray AK (2010) Automated leukocyte recognition using fuzzy divergence. Micron 41:840–846
14. Havrada JH, Charvat F (1967) Quantification methods of classification processes: concept of structural α-entropy. Kybernetika 3(1):30–35
15. Hwang CL, Yoon K (1981) Multiple attribute decision making—methods and applications. Springer, New York
16. Jahanshahloo GR, Hosseinzadeh LF, Izadikhah M (2006) Extension of TOPSIS method for decision making problems with fuzzy data. Appl Math Comput 181:1544–1551
17. Kapur JN (1997) Measures of fuzzy information. Mathematical Sciences Trust Society, New Delhi
18. Kullback S, Leibler RA (1951) On information and sufficiency. Ann Math Stat 22(1):79–86
19. Montes S, Couso I, Gil P, Bertoluzza C (2002) Divergence measure between fuzzy sets. Int J Approximate Reasoning 30:91–105
20. Pal NR, Pal SK (1989) Object background segmentation using new definition of entropy. IEE Proc Comput Digital Tech 136(4):248–295
21. Parkash O, Sharma PK, Kumar S (2006) Two new measures of fuzzy divergence and their properties. SQU J Sci 11:69–77
22. Poletti E, Zappelli F, Ruggeri A, Grisan E (2012) A review of thresholding strategies applied to human chromosome segmentation. Comput Methods Programs Biomed 108:679–688
23. Renyi A (1961) On measures of entropy and information. In: Proceeding of fourth Berkeley symposium on mathematics, statistics and probability, vol 1, pp 547–561
24. Stanujkic D, Magdalinovic N, Tojanovic S, Jovanovic R (2012) Extension of ratio system part of MOORA method for solving decision making problems with interval data. Informatica 23 (1):141–154
25. Verma R, Sharma BD (2011) On generalized exponential fuzzy entropy. World Acad Sci Eng Technol 69:1402–1405
26. Zadeh LA (1965) Fuzzy sets. Inf Control 8:338–353

Chapter 4
Sequence and Application of Inequalities Among Fuzzy Mean Difference Divergence Measures in Pattern Recognition

Fuzzy divergence measures and inequalities have recently been applied widely in the fuzzy comprehensive evaluation and information theory. In this chapter, we introduce a sequence of fuzzy mean difference divergence measures and establish the inequalities among them to explore the fuzzy inequalities. The advantage of establishing the inequalities is to make the computational work much simpler. The technique of inequalities provides a better comparison among fuzzy mean divergence measures.

In Sect. 4.1 we present preliminaries on fuzzy mean divergence measures and important results related to fuzzy measure that we use in our study. A sequence of fuzzy mean difference divergence measures and the essential properties for their validity are provided in Sect. 4.2. Section 4.3 provides some of inequalities among selected proposed fuzzy divergence measures. In the same section the proof of these inequalities is also presented. In Sect. 4.4 the applications of proposed fuzzy mean difference divergence measures in the context of pattern recognition have been presented using a numerical example. In the next section, it is shown that the proposed fuzzy mean difference divergence measures are well suited to use with linguistic variables. Finally, concluding remarks are drawn in Sect. 4.6.

4.1 Preliminaries on Fuzzy Mean Divergence Measures

The fuzzy mean divergence measures of Singh and Tomar [1] corresponding to seven geometrical mean measures given in Taneja [2] are presented in Table 4.1. We have the following Lemma in fuzzy context corresponding to the Lemma of Taneja [3]:

Lemma 4.1 *Let* $f : I \subset R^{+} \to R$ *be a convex and differentiable function satisfying* $f\left(\frac{1}{2}\right) = 0$. *Consider a function*

A. Ohlan and R. Ohlan, *Generalizations of Fuzzy Information Measures*,
DOI 10.1007/978-3-319-45928-8_4

Table 4.1 Fuzzy mean divergence measures

S. no.	Fuzzy mean divergence measure	Definition
1.	Fuzzy arithmetic mean measure	$A(A,B)=\sum_{i=1}^{n}\left(\frac{\mu_A(x_i)+\mu_B(x_i)}{2}+\frac{2-\mu_A(x_i)-\mu_B(x_i)}{2}\right)$
2.	Fuzzy geometric mean measure	$G(A,B)=\sum_{i=1}^{n}\left(\sqrt{\mu_A(x_i)\mu_B(x_i)}+\sqrt{(1-\mu_A(x_i))(1-\mu_B(x_i))}\right)$
3.	Fuzzy harmonic mean measure	$H(A,B)=\sum_{i=1}^{n}\left(\frac{2\mu_A(x_i)\mu_B(x_i)}{\mu_A(x_i)+\mu_B(x_i)}+\frac{2(1-\mu_A(x_i))(1-\mu_B(x_i))}{2-\mu_A(x_i)-\mu_B(x_i)}\right)$
4.	Fuzzy heronian mean measure	$N(A,B)=\sum_{i=1}^{n}\left(\frac{\mu_A(x_i)+\sqrt{\mu_A(x_i)\mu_B(x_i)}+\mu_B(x_i)}{3}+\frac{(1-\mu_A(x_i))+\sqrt{(1-\mu_A(x_i))(1-\mu_B(x_i))}+(1-\mu_B(x_i))}{3}\right)$
5.	Fuzzy contra-harmonic mean measure	$C(A,B)=\sum_{i=1}^{n}\left(\frac{\mu_A^2(x_i)+\mu_B^2(x_i)}{\mu_A(x_i)+\mu_B(x_i)}+\frac{(1-\mu_A(x_i))^2+(1-\mu_B(x_i))^2}{2-\mu_A(x_i)-\mu_B(x_i)}\right)$
6.	Fuzzy root-mean-square mean measure	$S(A,B)=\sum_{i=1}^{n}\left(\sqrt{\frac{\mu_A^2(x_i)+\mu_B^2(x_i)}{2}}+\sqrt{\frac{(1-\mu_A(x_i))^2+(1-\mu_B(x_i))^2}{2}}\right)$
7.	Fuzzy centroidal mean measure	$R(A,B)=\sum_{i=1}^{n}\left(\frac{2(\mu_A^2(x_i)+\mu_A(x_i)\mu_B(x_i)+\mu_B^2(x_i))}{3(\mu_A(x_i)+\mu_B(x_i))}+\frac{2\left((1-\mu_A(x_i))^2+(1-\mu_A(x_i))(1-\mu_B(x_i))+(1-\mu_B(x_i))^2\right)}{3(2-\mu_A(x_i)-\mu_B(x_i))}\right)$

$$\varphi_f(a,b)=af\left(\frac{b}{a}\right),\quad a,b>0,$$

Then the function $\varphi_f(a,b)$ is convex R_+^2. Additionally, if $f'(1/2)=0$, then the following inequality holds:

$$0\le\varphi_f(a,b)\le\left(\frac{b-a}{a}\right)\varphi_{f'}(a,b).$$

Lemma 4.2 (Schwarz's Lemma) *Let $f_1,f_2:I\subset R^+\to R$ be two convex functions satisfying the assumptions*

(i) $f_1\left(\frac{1}{2}\right)=f_1'\left(\frac{1}{2}\right)=0,\ f_2\left(\frac{1}{2}\right)=f_2'\left(\frac{1}{2}\right)=0;$

(ii) *f_1 and f_2 are twice differentiable in R^+;*

(iii) *there exist the real constants α,β such that $0\le\alpha<\beta$ and $\alpha\le\frac{f_1''(z)}{f_2''(z)}\le\beta$*

$f_2''(z)>0$, for all $z>0$ then we have the inequalities

$$\alpha\varphi_{f_2}(a,b)\le\varphi_{f_1}(a,b)\le\beta\varphi_{f_2}(a,b)$$

for all $a,b\in(0,1)$, where the function $\varphi_{(.)}(a,b)$ is defined as

$$\varphi_f(a,b) = af\left(\frac{b}{a}\right), \quad a,b > 0.$$

4.2 Sequence of Fuzzy Mean Difference Divergence Measures

Corresponding to the fuzzy mean divergence measures defined in Table 4.1 we propose a sequence of fuzzy mean difference divergence measures between fuzzy sets A and B of universe of discourse $X = \{x_1, x_2, \ldots, x_n\}$ having the membership values $\mu_A(x_i), \mu_B(x_i), i = 1, 2, \ldots, n$, as follows:

(i)

$$D_{CS}(A,B) = C(A,B) - S(A,B)$$
$$= \sum_{i=1}^{n} \left\{ \begin{array}{l} \left[\frac{\mu_A^2(x_i) + \mu_B^2(x_i)}{\mu_A(x_i) + \mu_B(x_i)} + \frac{(1-\mu_A(x_i))^2 + (1-\mu_B(x_i))^2}{2-\mu_A(x_i)-\mu_B(x_i)}\right] \\ -\left[\sqrt{\frac{\mu_A^2(x_i) + \mu_B^2(x_i)}{2}} + \sqrt{\frac{(1-\mu_A(x_i))^2 + (1-\mu_B(x_i))^2}{2}}\right] \end{array} \right\} \tag{4.1}$$

(ii)

$$D_{CN}(A,B) = C(A,B) - N(A,B)$$
$$= \sum_{i=1}^{n} \left\{ \begin{array}{l} \left[\frac{\mu_A^2(x_i) + \mu_B^2(x_i)}{\mu_A(x_i) + \mu_B(x_i)} + \frac{(1-\mu_A(x_i))^2 + (1-\mu_B(x_i))^2}{2-\mu_A(x_i)-\mu_B(x_i)}\right] \\ -\left[\begin{array}{l} \frac{\mu_A(x_i) + \sqrt{\mu_A(x_i)\mu_B(x_i)} + \mu_B(x_i)}{3} \\ + \frac{2-\mu_A(x_i)-\mu_B(x_i) + \sqrt{(1-\mu_A(x_i))(1-\mu_B(x_i))}}{3} \end{array}\right] \end{array} \right\} \tag{4.2}$$

(iii)

$$D_{CG}(A,B) = C(A,B) - G(A,B)$$
$$= \sum_{i=1}^{n} \left\{ \begin{array}{l} \left[\frac{\mu_A^2(x_i) + \mu_B^2(x_i)}{\mu_A(x_i) + \mu_B(x_i)} + \frac{(1-\mu_A(x_i))^2 + (1-\mu_B(x_i))^2}{2-\mu_A(x_i)-\mu_B(x_i)}\right] \\ -\left[\sqrt{\mu_A(x_i)\mu_B(x_i)} + \sqrt{(1-\mu_A(x_i))(1-\mu_B(x_i))}\right] \end{array} \right\} \tag{4.3}$$

(iv)

$$D_{CR}(A,B) = C(A,B) - R(A,B)$$
$$= \sum_{i=1}^{n} \left\{ \begin{array}{l} \left[\frac{\mu_A^2(x_i) + \mu_B^2(x_i)}{\mu_A(x_i) + \mu_B(x_i)} + \frac{(1-\mu_A(x_i))^2 + (1-\mu_B(x_i))^2}{2-\mu_A(x_i)-\mu_B(x_i)}\right] \\ -\left[\begin{array}{l} \frac{2(\mu_A^2(x_i) + \mu_A(x_i)\mu_B(x_i) + \mu_B^2(x_i))}{3(\mu_A(x_i) + \mu_B(x_i))} \\ + \frac{2((1-\mu_A(x_i))^2 + (1-\mu_A(x_i))(1-\mu_B(x_i)) + (1-\mu_B(x_i))^2)}{3(2-\mu_A(x_i)-\mu_B(x_i))} \end{array}\right] \end{array} \right\} \tag{4.4}$$

(v) $$D_{CA}(A,B) = C(A,B) - A(A,B)$$
$$= \sum_{i=1}^{n} \left\{ \begin{array}{c} \left[\frac{\mu_A^2(x_i) + \mu_B^2(x_i)}{\mu_A(x_i) + \mu_B(x_i)} + \frac{(1-\mu_A(x_i))^2 + (1-\mu_B(x_i))^2}{2-\mu_A(x_i)-\mu_B(x_i)}\right] \\ -\left[\frac{(\mu_A(x_i) + \mu_B(x_i))}{2} + \frac{2-\mu_A(x_i)-\mu_B(x_i)}{2}\right] \end{array} \right\} \tag{4.5}$$

(vi) $$D_{CH}(A,B) = C(A,B) - H(A,B)$$
$$= \sum_{i=1}^{n} \left\{ \begin{array}{c} \left[\frac{\mu_A^2(x_i) + \mu_B^2(x_i)}{\mu_A(x_i) + \mu_B(x_i)} + \frac{(1-\mu_A(x_i))^2 + (1-\mu_B(x_i))^2}{2-\mu_A(x_i)-\mu_B(x_i)}\right] \\ -\left[\frac{2\mu_A(x_i)\mu_B(x_i)}{\mu_A(x_i) + \mu_B(x_i)} + \frac{2(1-\mu_A(x_i))(1-\mu_B(x_i))}{2-\mu_A(x_i)-\mu_B(x_i)}\right] \end{array} \right\} \tag{4.6}$$

(vii) $$D_{SA}(A,B) = S(A,B) - A(A,B)$$
$$= \sum_{i=1}^{n} \left\{ \begin{array}{c} \left[\sqrt{\frac{(\mu_A^2(x_i) + \mu_B^2(x_i))}{2}} + \sqrt{\frac{((1-\mu_A(x_i))^2 + (1-\mu_B(x_i))^2)}{2}}\right] \\ -\left[\frac{(\mu_A(x_i) + \mu_B(x_i))}{2} + \frac{(2-\mu_A(x_i)-\mu_B(x_i))}{2}\right] \end{array} \right\} \tag{4.7}$$

(viii) $$D_{SN}(A,B) = S(A,B) - N(A,B)$$
$$= \sum_{i=1}^{n} \left\{ \begin{array}{c} \left[\sqrt{\frac{(\mu_A^2(x_i) + \mu_B^2(x_i))}{2}} + \sqrt{\frac{((1-\mu_A(x_i))^2 + (1-\mu_B(x_i))^2)}{2}}\right] \\ -\left[\begin{array}{c} \frac{\mu_A(x_i) + \sqrt{\mu_A(x_i)\mu_B(x_i)} + \mu_B(x_i)}{3} \\ + \frac{2-\mu_A(x_i)-\mu_B(x_i) + \sqrt{(1-\mu_A(x_i))(1-\mu_B(x_i))}}{3} \end{array}\right] \end{array} \right\} \tag{4.8}$$

(ix) $$D_{SG}(A,B) = S(A,B) - G(A,B)$$
$$= \sum_{i=1}^{n} \left\{ \begin{array}{c} \left[\sqrt{\frac{(\mu_A^2(x_i) + \mu_B^2(x_i))}{2}} + \sqrt{\frac{((1-\mu_A(x_i))^2 + (1-\mu_B(x_i))^2)}{2}}\right] \\ -\left[\sqrt{\mu_A(x_i)\mu_B(x_i)} + \sqrt{(1-\mu_A(x_i))(1-\mu_B(x_i))}\right] \end{array} \right\} \tag{4.9}$$

(x) $$D_{SH}(A,B) = S(A,B) - H(A,B)$$
$$= \sum_{i=1}^{n} \left\{ \begin{array}{c} \left[\sqrt{\frac{(\mu_A^2(x_i) + \mu_B^2(x_i))}{2}} + \sqrt{\frac{((1-\mu_A(x_i))^2 + (1-\mu_B(x_i))^2)}{2}}\right] \\ -\left[\frac{2\mu_A(x_i)\mu_B(x_i)}{\mu_A(x_i) + \mu_B(x_i)} + \frac{2(1-\mu_A(x_i))(1-\mu_B(x_i))}{2-\mu_A(x_i)-\mu_B(x_i)}\right] \end{array} \right\} \tag{4.10}$$

(xi) $D_{RA}(A,B) = R(A,B) - A(A,B)$

$$= \sum_{i=1}^{n} \left\{ \begin{array}{l} \left[\begin{array}{l} \frac{2(\mu_A^2(x_i)+\mu_A(x_i)\mu_B(x_i)+\mu_B^2(x_i))}{3(\mu_A(x_i)+\mu_B(x_i))} \\ \quad + \frac{2((1-\mu_A(x_i))^2+(1-\mu_A(x_i))(1-\mu_B(x_i))+(1-\mu_B(x_i))^2)}{3(2-\mu_A(x_i)-\mu_B(x_i))} \end{array} \right] \\ -\left[\frac{(\mu_A(x_i)+\mu_B(x_i))}{2} + \frac{(2-\mu_A(x_i)-\mu_B(x_i))}{2} \right] \end{array} \right\} \quad (4.11)$$

(xii) $D_{RN}(A,B) = R(A,B) - N(A,B)$

$$= \sum_{i=1}^{n} \left\{ \begin{array}{l} \left[\begin{array}{l} \frac{2(\mu_A^2(x_i)+\mu_A(x_i)\mu_B(x_i)+\mu_B^2(x_i))}{3(\mu_A(x_i)+\mu_B(x_i))} \\ + \frac{2((1-\mu_A(x_i))^2+(1-\mu_A(x_i))(1-\mu_B(x_i))+(1-\mu_B(x_i))^2)}{3(2-\mu_A(x_i)-\mu_B(x_i))} \end{array} \right] \\ -\left[\begin{array}{l} \frac{\mu_A(x_i)+\sqrt{\mu_A(x_i)\mu_B(x_i)}+\mu_B(x_i)}{3} \\ + \frac{2-\mu_A(x_i)-\mu_B(x_i)+\sqrt{(1-\mu_A(x_i))(1-\mu_B(x_i))}}{3} \end{array} \right] \end{array} \right\} \quad (4.12)$$

(xiii) $D_{RG}(A,B) = R(A,B) - G(A,B)$

$$= \sum_{i=1}^{n} \left\{ \begin{array}{l} \left[\begin{array}{l} \frac{2(\mu_A^2(x_i)+\mu_A(x_i)\mu_B(x_i)+\mu_B^2(x_i))}{3(\mu_A(x_i)+\mu_B(x_i))} \\ + \frac{2((1-\mu_A(x_i))^2+(1-\mu_A(x_i))(1-\mu_B(x_i))+(1-\mu_B(x_i))^2)}{3(2-\mu_A(x_i)-\mu_B(x_i))} \end{array} \right] \\ -\left[\sqrt{\mu_A(x_i)\mu_B(x_i)} + \sqrt{(1-\mu_A(x_i))(1-\mu_B(x_i))} \right] \end{array} \right\} \quad (4.13)$$

(xiv) $D_{RH}(A,B) = R(A,B) - H(A,B)$

$$= \sum_{i=1}^{n} \left\{ \begin{array}{l} \left[\begin{array}{l} \frac{2(\mu_A^2(x_i)+\mu_A(x_i)\mu_B(x_i)+\mu_B^2(x_i))}{3(\mu_A(x_i)+\mu_B(x_i))} \\ + \frac{2((1-\mu_A(x_i))^2+(1-\mu_A(x_i))(1-\mu_B(x_i))+(1-\mu_B(x_i))^2)}{3(2-\mu_A(x_i)-\mu_B(x_i))} \end{array} \right] \\ -\left[\frac{2\mu_A(x_i)\mu_B(x_i)}{\mu_A(x_i)+\mu_B(x_i)} + \frac{2(1-\mu_A(x_i))(1-\mu_B(x_i))}{2-\mu_A(x_i)-\mu_B(x_i)} \right] \end{array} \right\} \quad (4.14)$$

(xv) $D_{AN}(A,B) = A(A,B) - N(A,B)$

$$= \sum_{i=1}^{n} \left\{ \begin{array}{l} \left[\frac{(\mu_A(x_i)+\mu_B(x_i))}{2} + \frac{(2-\mu_A(x_i)-\mu_B(x_i))}{2} \right] \\ -\left[\begin{array}{l} \frac{\mu_A(x_i)+\sqrt{\mu_A(x_i)\mu_B(x_i)}+\mu_B(x_i)}{3} \\ + \frac{2-\mu_A(x_i)-\mu_B(x_i)+\sqrt{(1-\mu_A(x_i))(1-\mu_B(x_i))}}{3} \end{array} \right] \end{array} \right\} \quad (4.15)$$

(xvi)
$$D_{AG}(A,B) = A(A,B) - G(A,B) = \sum_{i=1}^{n} \left\{ \begin{array}{l} \left[\frac{(\mu_A(x_i)+\mu_B(x_i))}{2} + \frac{(2-\mu_A(x_i)-\mu_B(x_i))}{2}\right] \\ -\left[\sqrt{\mu_A(x_i)\mu_B(x_i)} + \sqrt{(1-\mu_A(x_i))(1-\mu_B(x_i))}\right] \end{array} \right\} \tag{4.16}$$

(xvii)
$$D_{AH}(A,B) = A(A,B) - H(A,B) \tag{4.17}$$
$$= \sum_{i=1}^{n} \left\{ \begin{array}{l} \left[\frac{(\mu_A(x_i)+\mu_B(x_i))}{2} + \frac{(2-\mu_A(x_i)-\mu_B(x_i))}{2}\right] \\ -\left[\frac{2\mu_A(x_i)\mu_B(x_i)}{\mu_A(x_i)+\mu_B(x_i)} + \frac{2(1-\mu_A(x_i))(1-\mu_B(x_i))}{2-\mu_A(x_i)-\mu_B(x_i)}\right] \end{array} \right\}$$

(xviii)
$$D_{NG}(A,B) = N(A,B) - G(A,B) = \sum_{i=1}^{n} \left\{ \begin{array}{l} \left[\begin{array}{l} \frac{\mu_A(x_i)+\sqrt{\mu_A(x_i)\mu_B(x_i)}+\mu_B(x_i)}{3} \\ + \frac{2-\mu_A(x_i)-\mu_B(x_i)+\sqrt{(1-\mu_A(x_i))(1-\mu_B(x_i))}}{3} \end{array}\right] \\ -\left[\sqrt{\mu_A(x_i)\mu_B(x_i)} + \sqrt{(1-\mu_A(x_i))(1-\mu_B(x_i))}\right] \end{array} \right\} \tag{4.18}$$

Theorem 4.1 *All the proposed measures from Eqs. (4.1) to (4.18) are valid measures of fuzzy mean difference divergence.*

Proof

(a) **Nonnegativity**
From one of inequalities given in Taneja [2], for two fuzzy sets A and B we have $H(A,B) \le G(A,B) \le N(A,B) \le A(A,B) \le R(A,B) \le S(A,B) \le C(A,B)$. Hence, the condition of nonnegativity of measures from Eqs. (4.1) to (4.18) is proved.

(b) Also clearly, $D_{A_1B_1}(A,A) = 0$ for all measures from Eqs. (4.1) to (4.18) where A_1 and B_1 belong to the fuzzy mean divergence measures given in Table 4.1.

(c) **Convexity**
We now shall prove the condition of convexity of measures for Eqs. (4.1)–(4.18) with the help of Lemma 4.1.

For simplicity, let us write $D_{A_1B_1} = bf_{A_1B_1}$ where $f_{A_1B_1}(z) = f_{A_1}(z) - f_{B_1}(z)$ with $A_1 \ge B_1$.

Let us take $\mu_A = z \Rightarrow \mu_B = 1 - z$. So, corresponding to measures of Eqs. (4.1)–(4.18) we have the following generating functions:

1.
$$f_{CS}(z) = 2\left[z^2 + (1-z)^2 - \sqrt{\frac{z^2+(1-z)^2}{2}}\right] \tag{4.19}$$

2. $$f_{CN}(z) = 2\left[z^2 + (1-z)^2 - \frac{1+\sqrt{z(1-z)}}{3}\right] \tag{4.20}$$

3. $$f_{CG}(z) = 2\left[z^2 + (1-z)^2 - \sqrt{z(1-z)}\right] \tag{4.21}$$

4. $$f_{CR}(z) = 2\left[z^2 + (1-z)^2 - \frac{2(z^2 + (1-z)^2 + z(1-z))}{3}\right] \tag{4.22}$$

5. $$f_{CA}(z) = 2\left[z^2 + (1-z)^2 - \frac{z+(1-z)}{2}\right] \tag{4.23}$$

6. $$f_{CH}(z) = 2\left[z^2 + (1-z)^2 - 2z(1-z)\right] \tag{4.24}$$

7. $$f_{SA}(z) = 2\left[\sqrt{\frac{z^2 + (1-z)^2}{2}} - \frac{z+(1-z)}{2}\right] \tag{4.25}$$

8. $$f_{SN}(z) = 2\left[\sqrt{\frac{z^2 + (1-z)^2}{2}} - \frac{z+\sqrt{z(1-z)}+(1-z)}{3}\right] \tag{4.26}$$

9. $$f_{SG}(z) = 2\left[\sqrt{\frac{z^2 + (1-z)^2}{2}} - \sqrt{z(1-z)}\right] \tag{4.27}$$

10. $$f_{SH}(z) = 2\left[\sqrt{\frac{z^2 + (1-z)^2}{2}} - 2z(1-z)\right] \tag{4.28}$$

11. $$f_{RA}(z) = 2\left[\frac{2(z^2 + (1-z)^2 + z(1-z))}{3} - \frac{z+(1-z)}{2}\right] \tag{4.29}$$

12. $$f_{RN}(z) = 2\left[\frac{2(z^2 + (1-z)^2 + z(1-z))}{3} - \frac{1+\sqrt{z(1-z)}}{3}\right] \tag{4.30}$$

13. $$f_{RG}(z) = 2\left[\frac{2(z^2 + (1-z)^2 + z(1-z))}{3} - \sqrt{z(1-z)}\right] \tag{4.31}$$

14.
$$f_{RH}(z) = 2\left[\frac{2(z^2+(1-z)^2+z(1-z))}{3} - 2z(1-z)\right] \tag{4.32}$$

15.
$$f_{AN}(z) = 2\left[\frac{z+(1-z)}{2} - \frac{1+\sqrt{z(1-z)}}{3}\right] \tag{4.33}$$

16.
$$f_{AG}(z) = 2\left[\frac{z+(1-z)}{2} - \sqrt{z(1-z)}\right] \tag{4.34}$$

17.
$$f_{AH}(z) = 2\left[\frac{z+(1-z)}{2} - 2z(1-z)\right] \tag{4.35}$$

18.
$$f_{NG}(z) = 2\left[\frac{1+\sqrt{z(1-z)}}{3} - \sqrt{z(1-z)}\right] \tag{4.36}$$

Now in all the cases from Eqs. (4.19) to (4.36), we can easily check that $f_{A_1B_1}\left(\frac{1}{2}\right) = f_{A_1}\left(\frac{1}{2}\right) - f_{B_1}\left(\frac{1}{2}\right) = \frac{1}{2} - \frac{1}{2} = 0$. It is understood that $z \in [0, 1]$.

The first- and second-order derivatives of the functions Eqs. (4.19)–(4.36) are as follows:

1.
$$\begin{aligned} f'_{CS}(z) &= 4(2z-1) - \frac{2(2z-1)}{\sqrt{2(z^2+(1-z)^2)}}, \\ f''_{CS}(z) &= 8 - \frac{4}{\left(2(z^2+(1-z)^2)\right)^{3/2}} > 0 \end{aligned} \tag{4.37}$$

2.
$$\begin{aligned} f'_{CN}(z) &= 4(2z-1) + \frac{(2z-1)}{3\sqrt{z-z^2}}, \\ f''_{CN}(z) &= 8 + \frac{4}{6(z-z^2)^{3/2}} > 0 \end{aligned} \tag{4.38}$$

3.
$$\begin{aligned} f'_{CG}(z) &= 4(2z-1) + \frac{(2z-1)}{\sqrt{z-z^2}}, \\ f''_{CG}(z) &= 8 + \frac{4}{2(z-z^2)^{3/2}} > 0 \end{aligned} \tag{4.39}$$

4.
$$f'_{CR}(z) = \frac{8(2z-1)}{3}, \quad f''_{CR}(z) = \frac{16}{3} > 0 \tag{4.40}$$

5. $$f'_{CA}(z) = 4(2z-1), \quad f''_{CA}(z) = 8 > 0 \tag{4.41}$$

6. $$f'_{CH}(z) = 8(2z-1), \quad f''_{CH}(z) = 16 > 0 \tag{4.42}$$

7. $$f'_{SA}(z) = \frac{2(2z-1)}{\sqrt{2(z^2+(1-z)^2)}}, \quad f''_{SA}(z) = \frac{\sqrt{2}}{\left(z^2+(1-z)^2\right)^{3/2}} > 0 \tag{4.43}$$

8. $$\begin{aligned} f'_{SN}(z) &= \frac{2(2z-1)}{\sqrt{2(z^2+(1-z)^2)}} + \frac{(2z-1)}{3\sqrt{z-z^2}}, \\ f''_{SN}(z) &= \frac{\sqrt{2}}{\left(z^2+(1-z)^2\right)^{3/2}} + \frac{4}{6(z-z^2)^{3/2}} > 0 \end{aligned} \tag{4.44}$$

9. $$\begin{aligned} f'_{SG}(z) &= \frac{2(2z-1)}{\sqrt{2(z^2+(1-z)^2)}} + \frac{(2z-1)}{\sqrt{z-z^2}}, \\ f''_{SG}(z) &= \frac{4}{\left(2(z^2+(1-z)^2)\right)^{3/2}} + \frac{1}{2(z-z^2)^{3/2}} > 0 \end{aligned} \tag{4.45}$$

10. $$\begin{aligned} f'_{SH}(z) &= \frac{4(2z-1)}{\sqrt{2(z^2+(1-z)^2)}} + 4(2z-1), \\ f''_{SH}(z) &= 8 + \frac{\sqrt{2}}{\left((z^2+(1-z)^2)\right)^{3/2}} > 0 \end{aligned} \tag{4.46}$$

11. $$f'_{RA}(z) = \frac{4(2z-1)}{3}, \quad f''_{RA}(z) = \frac{8}{3} > 0 \tag{4.47}$$

12. $$\begin{aligned} f'_{RN}(z) &= \frac{4(2z-1)}{3} + \frac{(2z-1)}{3\sqrt{z-z^2}}, \\ f''_{RN}(z) &= \frac{8}{3} + \frac{1}{3}\left[\frac{z^2+(1-z)^2}{(z-z^2)^{3/2}}\right] > 0 \end{aligned} \tag{4.48}$$

13. $$\begin{aligned} f'_{RG}(z) &= \frac{4(2z-1)}{3} + \frac{(2z-1)}{\sqrt{z-z^2}}, \\ f''_{RG}(z) &= \frac{4}{3} + \frac{1}{2(z-z^2)^{3/2}} > 0 \end{aligned} \tag{4.49}$$

14. $$f'_{RH}(z) = \frac{16(2z-1)}{3}, \quad f''_{RH}(z) = \frac{32}{3} > 0 \tag{4.50}$$

15. $$f'_{AN}(z) = \frac{(2z-1)}{3\sqrt{z-z^2}}, \quad f''_{AN}(z) = \frac{1}{6(z-z^2)^{3/2}} > 0 \tag{4.51}$$

16. $$f'_{AG}(z) = \frac{(2z-1)}{\sqrt{z-z^2}}, \quad f''_{AG}(z) = \frac{1}{(z-z^2)^{3/2}} > 0 \tag{4.52}$$

17. $$f'_{AH}(z) = 4(2z-1), \quad f''_{AH}(z) = 8 > 0 \tag{4.53}$$

18. $$f'_{NG}(z) = \frac{2(2z-1)}{3\sqrt{z-z^2}}, \quad f''_{NG}(z) = \frac{2(z^2+(1-z)^2)}{3(z-z^2)^{3/2}} > 0 \tag{4.54}$$

We see that in the all cases second-order derivative is positive and satisfies $f'_{A_1B_1}\left(\frac{1}{2}\right) = 0$ for all $z \in [0,1]$. Thus, according to Lemma 4.1 and Eqs. (4.39)–(4.54), we get the convexity of the measures for Eqs. (4.1)–(4.18).

Hence in view of the definition of fuzzy divergence measure of Bhandari and Pal [4], all the defined measures for Eqs. (4.1)–(4.18) are valid measures of fuzzy mean difference divergence. Moreover, we can easily check for measures in Eqs. (4.1)–(4.18) such that $D_{A_1B_1}(A,B) = D_{A_1B_1}(B,A)$ where A_1 and B_1 belong to the fuzzy mean divergence measures given in Table 4.1. Hence all the defined measures for Eqs. (4.1)–(4.18) are valid measures of fuzzy symmetric mean difference divergence.

4.3 Inequalities Among Fuzzy Mean Difference Divergence Measures

Theorem 4.2 *The fuzzy mean difference divergence measures defined in Eqs. (4.1)–(4.18) admit the following inequalities:*

$$D_{SA} \le \left\{ \begin{array}{c} \frac{3}{4}D_{SN} \\ \frac{1}{3}D_{SH} \le \frac{3}{4}D_{CR} \end{array} \right\} \le \left\{ \begin{array}{c} \frac{3}{7}D_{CN} \le \left\{ \begin{array}{c} D_{CS} \\ \frac{1}{3}D_{CG} \le \frac{3}{5}D_{RG} \end{array} \right. \\ \frac{1}{2}D_{SG} \le \frac{3}{5}D_{RG} \end{array} \right\} \le 3D_{AN}$$

That is, we have the following inequalities:

(i) $$D_{SA} \le \frac{3}{4}D_{SN} \le \frac{3}{7}D_{CN} \le D_{CS} \le 3D_{AN},$$

(ii) $$D_{SA} \le \frac{1}{3}D_{SH} \le \frac{3}{4}D_{CR} \le \frac{1}{2}D_{SG} \le \frac{3}{5}D_{RG} \le 3D_{AN},$$

(iii) $$D_{SA} \le \frac{1}{3} D_{SH} \le \frac{3}{4} D_{CR} \le \frac{3}{7} D_{CN} \le \frac{1}{3} D_{CG} \le \frac{3}{5} D_{RG} \le 3 D_{AN},$$

(iv) $$D_{SA} \le \frac{3}{4} D_{SN} \le \frac{1}{2} D_{SG} \le \frac{3}{5} D_{RG} \le 3 D_{AN}.$$

Proof The proof of the above theorem is based on Lemma 4.2 and is given in parts in the following propositions.

Proposition 4.1 *We have* $D_{SA} \le \frac{3}{4} D_{SN}$

Proof Let us consider the function

$$g_{SA\_SN}(z) = \frac{f''_{SA}(z)}{f''_{SN}(z)} = \frac{6\sqrt{2}(z - z^2)^{3/2}}{6\sqrt{2}(z - z^2)^{3/2} + \left((z^2 + (1 - z)^2)\right)^{3/2}}$$

This gives

$$g'_{SA\_SN}(z) = \frac{3\sqrt{2}(2z - 1)(z - z^2)^{1/2}(2z^2 - 2z + 1)^{1/2}(4z^2 - 4z - 1)}{\left[6\sqrt{2}(z - z^2)^{3/2} + (2z^2 - 2z + 1)^{3/2}\right]^2} \begin{cases} > 0 & \text{for } z < 1/2 \\ < 0 & \text{for } z > 1/2 \end{cases}$$

We have

$$\beta = \sup_{z \in [0,1]} g_{SA\_SN}(z) = g_{SA\_SN}\left(\frac{1}{2}\right) = \frac{3}{4} \tag{4.55}$$

Applying Lemma 4.2 for the difference of fuzzy means $D_{SA}(A, B)$ and $D_{SN}(A, B)$ and using Eq. (4.55) we get

$$D_{SA} \le \frac{3}{4} D_{SN}.$$

Proposition 4.2 *We have* $D_{SA} \le \frac{1}{3} D_{SH}$

Proof Let us consider the function

$$g_{SA\_SH}(z) = \frac{f''_{SA}(z)}{f''_{SH}(z)} = \frac{\sqrt{2}}{8(2z^2 - 2z + 1)^{3/2} + \sqrt{2}}$$

This gives

$$g'_{SA\_SH}(z) = -\frac{12\sqrt{2}(2z^2 - 2z + 1)^{1/2}(4z - 2)}{\left[8(2z^2 - 2z + 1)^{3/2} + \sqrt{2}\right]^2} \begin{cases} > 0 & \text{for } z < 1/2 \\ < 0 & \text{for } z > 1/2 \end{cases}$$

We have

$$\beta = \sup_{z\in[0,1]} g_{SA\_SH}(z) = g_{SA\_SH}\left(\frac{1}{2}\right) = \frac{1}{3} \tag{4.56}$$

Applying Lemma 4.2 for the difference of fuzzy means $D_{SA}(A,B)$ and $D_{SH}(A,B)$ and using Eq. (4.56) we get

$$D_{SA} \le \frac{1}{3} D_{SH}.$$

Proposition 4.3 *We have* $D_{SH} \le \frac{9}{4} D_{CR}$

Proof Let us consider the function

$$g_{SH\_CR}(z) = \frac{f''_{SH}(z)}{f''_{CR}(z)} = \frac{24(2z^2-2z+1)^{3/2}+3\sqrt{2}}{16(2z^2-2z+1)^{3/2}}$$

This gives

$$g'_{SH\_CR}(z) = -\frac{46\sqrt{2}(2z-1)}{49(2z^2-2z+1)^{5/2}} \begin{cases} >0 & \text{for } z<1/2 \\ <0 & \text{for } z>1/2 \end{cases}$$

And we have

$$\beta = \sup_{z\in[0,1]} g_{SH\_CR}(z) = g_{SH\_CR}\left(\frac{1}{2}\right) = \frac{9}{4} \tag{4.57}$$

Applying Lemma 4.2 for the difference of fuzzy means $D_{SH}(A,B)$ and $D_{CR}(A,B)$ and using Eq. (4.57) we get

$$D_{SH} \le \frac{9}{4} D_{CR}.$$

Proposition 4.4 *We have* $D_{CR} \le \frac{4}{7} D_{CN}$

Proof Let us consider the function

$$g_{CR\_CN}(z) = \frac{f''_{CR}(z)}{f''_{CN}(z)} = \frac{32(z-z^2)^{3/2}}{48(z-z^2)^{3/2}+1}$$

This gives

$$g'_{CR\_CN}(z) = \frac{48(z-z^2)^{1/2}(1-2z)}{\left[48(z-z^2)^{3/2}+1\right]^2}\begin{cases} >0 & \text{for } z<1/2 \\ <0 & \text{for } z>1/2 \end{cases}$$

We have

$$\beta = \sup_{z\in[0,1]} g_{CR\_CN}(z) = g_{CR\_CN}\left(\frac{1}{2}\right) = \frac{4}{7} \qquad (4.58)$$

Applying Lemma 4.2 for the difference of fuzzy means $D_{CR}(A,B)$ and $D_{CN}(A,B)$ and using Eq. (4.58) we get

$$D_{CR} \leq \frac{4}{7} D_{CN}.$$

Proposition 4.5 *We have* $D_{CR} \leq \frac{2}{3} D_{SG}$

Proof Let us consider the function

$$g_{CR\_CN}(z) = \frac{f''_{CR}(z)}{f''_{SG}(z)} = \frac{16(4z^2-4z+2)^{3/2}(z-z^2)^{3/2}}{24(z-z^2)^{3/2}+3(4z^2-4z+2)^{3/2}}$$

This gives

$$g'_{CR\_SG}(z) = \frac{8(2z-1)(z-z^2)^{1/2}(4z^2-4z+2)^{1/2}\left[32(z-z^2)^{5/2}-(4z^2-4z+2)^{5/2}\right]}{\left[8(z-z^2)^{3/2}+(4z^2-4z+2)^{3/2}\right]^2}$$
$$\begin{cases} >0 & \text{for } z<1/2 \\ <0 & \text{for } z>1/2 \end{cases}$$

We have

$$\beta = \sup_{z\in[0,1]} g_{CR\_SG}(z) = g_{CR\_SG}\left(\frac{1}{2}\right) = \frac{2}{3} \qquad (4.59)$$

Applying Lemma 4.2 for the difference of fuzzy means $D_{CR}(A,B)$ and $D_{SG}(A,B)$ and using Eq. (4.59) we get

$$D_{CR} \leq \frac{2}{3} D_{SG}.$$

Proposition 4.6 *We have* $D_{SN} \leq \frac{4}{7} D_{CN}$

Proof Let us consider the function

$$g_{SN\_CN}(z) = \frac{f''_{SN}(z)}{f''_{CN}(z)} = \frac{24(z-z^2)^{3/2} + (4z^2-4z+2)^{3/2}}{[48(z-z^2)^{3/2}+1](4z^2-4z+2)^{3/2}}$$

This gives

$$g'_{SN\_CN}(z) = \frac{72(z-z^2)^{1/2}(4z^2-4z+2)^{1/2}(1-2z)\left[1+96(z-z^2)^{5/2}-(4z^2-4z+2)^{5/2}\right]}{\left(48(z-z^2)^{3/2}+1\right)^2(4z^2-4z+2)^3}$$
$$\begin{cases} >0 & \text{for } z<1/2 \\ <0 & \text{for } z>1/2 \end{cases}$$

We have

$$\beta = \sup_{z\in[0,1]} g_{SN\_CN}(z) = g_{SN\_CN}\left(\frac{1}{2}\right) = \frac{4}{7} \tag{4.60}$$

Applying Lemma 4.2 for the difference of fuzzy means $D_{SN}(A,B)$ and $D_{CN}(A,B)$ and using Eq. (4.60) we get

$$D_{SN} \leq \frac{4}{7} D_{CN}.$$

Proposition 4.7 *We have* $D_{SN} \leq \frac{2}{3} D_{SG}$

Proof Let us consider the function

$$g_{SN\_SG}(z) = \frac{f''_{SN}(z)}{f''_{SG}(z)} = \frac{24(z-z^2)^{3/2} + (4z^2-4z+2)^{3/2}}{24(z-z^2)^{3/2}+3(4z^2-4z+2)^{3/2}}$$

This gives

$$g'_{SN\_SG}(z) = \frac{144(z-z^2)^{1/2}(4z^2-4z+2)^{1/2}(1-2z)}{\left[24(z-z^2)^{3/2}+3(4z^2-4z+2)^{3/2}\right]^2} \begin{cases} >0 & \text{for } z<1/2 \\ <0 & \text{for } z>1/2 \end{cases}$$

We have

$$\beta = \sup_{z\in[0,1]} g_{SN\_SG}(z) = g_{SN\_SG}\left(\frac{1}{2}\right) = \frac{2}{3}. \tag{4.61}$$

Applying Lemma 4.2 for the difference of fuzzy means $D_{SN}(A,B)$ and $D_{SG}(A,B)$ and using Eq. (4.61) we get

$$D_{SN} \leq \frac{2}{3} D_{SG}.$$

Proposition 4.8 *We have* $D_{CN} \leq \frac{7}{3} D_{CS}$

Proof Let us consider the function

$$g_{CN\_CS}(z) = \frac{f''_{CN}(z)}{f''_{CS}(z)} = \left[8 + \frac{1}{6(z-z^2)^{3/2}}\right]\left[8 - \frac{4}{(4z^2-4z+2)^{3/2}}\right]^{-1}$$

This gives

$$g'_{CN\_CS}(z) = \frac{(2z-1)(4z^2-4z+2)^{1/2}\{(4z^2-4z+2)[8(4z^2-4z+2)^{3/2}-4]-16(z-z^2)[48(z-z^2)^{3/2}+1]\}}{4(z-z^2)^{5/2}[8(4z^2-4z+2)^{3/2}-4]^2}$$

$$\begin{cases} >0 & \text{for } z<1/2 \\ <0 & \text{for } z>1/2 \end{cases}$$

We have

$$\beta = \sup_{z\in[0,1]} g_{CN\_CS}(z) = g_{CN\_CS}\left(\frac{1}{2}\right) = \frac{7}{3} \tag{4.62}$$

Applying Lemma 4.2 for the difference of fuzzy means $D_{CN}(A,B)$ and $D_{CS}(A,B)$ and using Eq. (4.62) we get

$$D_{CN} \leq \frac{7}{3} D_{CS}.$$

Proposition 4.9 *We have* $D_{CS} \leq 3D_{AN}$

Proof Let us consider the function

$$g_{CS\_AN}(z) = \frac{f''_{CS}(z)}{f''_{AN}(z)} = \left[\frac{8(4z^2-4z+2)^{3/2}-4}{(4z^2-4z+2)^{3/2}}\right]\left[6(z-z^2)^{3/2}\right]$$

This gives

$$g'_{CS\_AN}(z) = \frac{36(2z-1)(z-z^2)\left\{4(z-z^2)^{1/2}-(4z^2-4z+2)\left[2(4z^2-4z+2)^{3/2}-1\right]\right\}}{(4z^2-4z+2)^{5/2}} \begin{cases} >0 & \text{for } z<1/2 \\ <0 & \text{for } z>1/2 \end{cases}$$

We have

$$\beta = \sup_{z \in [0,1]} g_{CS\_AN}(z) = g_{CS\_AN}\left(\frac{1}{2}\right) = 3 \tag{4.63}$$

Applying Lemma 4.2 for the difference of fuzzy means $D_{CS}(A,B)$ and $D_{AN}(A,B)$ and using Eq. (4.63) we get

$$D_{CS} \leq 3D_{AN}.$$

Proposition 4.10 *We have* $D_{CN} \leq \frac{7}{9} D_{CG}$

Proof Let us consider the function

$$g_{CN\_CG}(z) = \frac{f''_{CN}(z)}{f''_{CG}(z)} = 1 - 2[48(z - z^2)^{3/2} + 3]^{-1}$$

This gives

$$g'_{CN\_CG}(z) = \frac{16(1-2z)(z-z^2)^{1/2}}{\left[16(z-z^2)^{3/2}+1\right]^2} \begin{cases} > 0 & \text{for } z < 1/2 \\ < 0 & \text{for } z > 1/2 \end{cases}$$

We have

$$\beta = \sup_{z \in [0,1]} g_{CN\_CG}(z) = g_{CN\_CG}\left(\frac{1}{2}\right) = \frac{7}{9} \tag{4.64}$$

Applying Lemma 4.2 for the difference of fuzzy means $D_{CN}(A,B)$ and $D_{CG}(A,B)$ and using Eq. (4.64) we get

$$D_{CN} \leq \frac{7}{9} D_{CG}.$$

Proposition 4.11 *We have* $D_{SG} \leq \frac{6}{5} D_{RG}$

Proof Let us consider the function

$$g_{SG\_RG}(z) = \frac{f''_{SG}(z)}{f''_{RG}(z)} = \frac{24(z-z^2)^{3/2} + 3(4z^2 - 4z + 2)^{3/2}}{(4z^2 - 4z + 2)^{3/2}[8(z-z^2)^{3/2} + 3]}$$

This gives

$$g'_{SG\_RG}(z) = \frac{9(2z-1)\left\{\begin{array}{c}[(4z^2-4z+2)^{5/2}-32(z-z^2)^{5/2}][8(z-z^2)^{3/2}+3]\\ -3[8(z-z^2)^{3/2}+(4z^2-4z+2)^{3/2}](4z^2-4z+2)\end{array}\right\}}{2(4z^2-4z+2)^{5/2}(z-z^2)\left[8(z-z^2)^{3/2}+3\right]^2}$$

$$g'_{SG\_RG}(z)\begin{cases} >0 & \text{for } z<1/2 \\ <0 & \text{for } z>1/2 \end{cases}$$

We have

$$\beta = \sup_{z\in[0,1]} g_{SG\_RG}(z) = g_{SG\_RG}\left(\frac{1}{2}\right) = \frac{6}{5} \tag{4.65}$$

Applying Lemma 4.2 for the difference of fuzzy means $D_{SG}(A,B)$ and $D_{RG}(A,B)$ and using Eq. (4.65) we get

$$D_{SG} \leq \frac{6}{5} D_{RG}.$$

Proposition 4.12 *We have* $D_{CG} \leq \frac{9}{5} D_{RG}$

Proof Let us consider the function

$$g_{CG\_RG}(z) = \frac{f''_{CG}(z)}{f''_{RG}(z)} = 6 - 45[8(z-z^2)^{3/2}+3]^{-1}$$

This gives

$$g'_{CG\_RG}(z) = \frac{540(z-z^2)^{1/2}(1-2z)}{8(z-z^2)^{3/2}+3}\begin{cases} >0 & \text{for } z<1/2 \\ <0 & \text{for } z>1/2 \end{cases}$$

We have

$$\beta = \sup_{z\in[0,1]} g_{CG\_RG}(z) = g_{CG\_RG}\left(\frac{1}{2}\right) = \frac{9}{5} \tag{4.66}$$

Applying Lemma 4.2 for the difference of fuzzy means $D_{CG}(A,B)$ and $D_{RG}(A,B)$ and using Eq. (4.66) we get

$$D_{CG} \leq \frac{9}{5} D_{RG}.$$

Proposition 4.13 *We have* $D_{RG} \leq 5D_{AN}$

Proof Let us consider the function

$$g_{RG\_AN}(z) = \frac{f''_{RG}(z)}{f''_{AN}(z)} = 8(z - z^2)^{3/2} + 3$$

This gives

$$g'_{RG\_AN}(z) = 8(z - z^2)^{3/2} + 3 \begin{cases} > 0 & \text{for } z < 1/2 \\ < 0 & \text{for } z > 1/2 \end{cases}$$

We have

$$\beta = \sup_{z \in [0,1]} g_{RG\_AN}(z) = g_{RG\_AN}\left(\frac{1}{2}\right) = 5 \tag{4.67}$$

Applying Lemma 4.2 for the difference of fuzzy means $D_{RG}(A, B)$ and $D_{AN}(A, B)$ and using Eq. (4.67) we get

$$D_{RG} \leq 5D_{AN}.$$

4.4 Application of Fuzzy Mean Difference Divergence Measures to Pattern Recognition

We now present the application of the proposed fuzzy mean difference divergence measures in the context of pattern recognition. Next, an example related to pattern recognition is given to demonstrate the results obtained for the fuzzy mean difference divergence measures for Eqs. (4.1)–(4.18).

In order to demonstrate the application of the introduced fuzzy mean difference divergence measures to pattern recognition, suppose that we are given three known patterns P_1, P_2 and P_3 which have classifications C_1, C_2 and C_3 respectively. These patterns are represented by the following fuzzy sets in the universe of discourse $X = \{x_1, x_2, x_3, x_4\}$:

$$P_1 = \{\langle x_1, 0.5\rangle, \langle x_2, 0.6\rangle, \langle x_3, 0.2\rangle, \langle x_4, 0.3\rangle\}$$
$$P_2 = \{\langle x_1, 0.8\rangle, \langle x_2, 0.7\rangle, \langle x_3, 0.3\rangle, \langle x_4, 0.4\rangle\}$$
$$P_3 = \{\langle x_1, 0.7\rangle, \langle x_2, 0.5\rangle, \langle x_3, 0.1\rangle, \langle x_4, 0.7\rangle\}$$

Given an unknown pattern Q, represented by the fuzzy set

Table 4.2 Computed values of fuzzy mean difference divergence measures $D_{AB}(P_k, Q)$ with $k = \{1, 2, 3\}$

	P_1	P_2	P_3
Q	$0.2741^{(6)}$	$0.2877^{(6)}$	$0.1385^{(6)}$
	$0.5825^{(7)}$	$0.6430^{(7)}$	$0.3127^{(7)}$
	$0.8002^{(8)}$	$0.8400^{(8)}$	$0.4064^{(8)}$
	$0.3423^{(9)}$	$0.3631^{(9)}$	$0.1772^{(9)}$
	$0.5135^{(10)}$	$0.5446^{(10)}$	$0.2659^{(10)}$
	$1.0270^{(11)}$	$1.0892^{(11)}$	$0.5318^{(11)}$
	$0.2385^{(12)}$	$0.2569^{(12)}$	$0.1274^{(12)}$
	$0.3340^{(13)}$	$0.3553^{(13)}$	$0.1742^{(13)}$
	$0.5252^{(14)}$	$0.5523^{(14)}$	$0.2679^{(14)}$
	$0.7520^{(15)}$	$0.8015^{(15)}$	$0.3933^{(15)}$
	$0.1712^{(16)}$	$0.1815^{(16)}$	$0.0887^{(16)}$
	$0.2667^{(17)}$	$0.2799^{(17)}$	$0.1355^{(17)}$
	$0.4579^{(18)}$	$0.4769^{(18)}$	$0.2292^{(18)}$
	$0.6847^{(19)}$	$0.7261^{(19)}$	$0.3546^{(19)}$
	$0.0955^{(20)}$	$0.0984^{(20)}$	$0.0468^{(20)}$
	$0.2867^{(21)}$	$0.2954^{(21)}$	$0.1405^{(21)}$
	$0.5135^{(22)}$	$0.5446^{(22)}$	$0.2659^{(22)}$
	$0.1912^{(23)}$	$0.1970^{(23)}$	$0.0937^{(23)}$

$$Q = \{\langle x_1, 0.5\rangle, \langle x_2, 0.3\rangle, \langle x_3, 0.4\rangle, \langle x_4, 0.9\rangle\}.$$

For convenience we use the notation $*^{(i)}$ in Table 4.2 to present the divergence value computed from equation *i*.

Our aim here is to classify Q to one of the classes C_1, C_2 and C_3. According to the principle of minimum divergence information between fuzzy sets, the process of assigning Q to C_{k^*} is described by $k^* = \arg\min_k\{D_{AB}(P_k, Q)\}$.

Table 4.2 (above) presents $D_{AB}(P_k, Q)$, $k = \{1, 2, 3\}$. It is observed that Q has been classified to C_3 correctly.

4.5 Numerical Example

Let us establish that the proposed fuzzy mean difference divergence measures Eqs. (4.1)–(4.18) are reliable in applications with compound linguistic variables.

Example Let $F = \{(x, \mu_F(x))/x \in X\}$ be a fuzzy set in X. Tomar and Ohlan [5, 6] defined for any positive real number n, from the operation of power of a fuzzy set

$$F^n = \{(x, [\mu_F(x)]^n)/x \in X\}.$$

Using the above operation, the concentration and dilation of a fuzzy set *F* are as follows:

Concentration: CON (*F*) = F^2,
Dilation: DIL (*F*) = $F^{1/2}$.
CON (*F*) and DIL (*F*) are treated as 'very (*F*)' and 'more or less (*F*)', respectively.

We consider *F* in $X = \{x_1, x_2, x_3, x_4, x_5\}$ defined as

$$F = \{(x_1, 0.3), (x_2, 0.6), (x_3, 0.9), (x_4, 0.5), (x_5, 0.1)\}.$$

By taking into account the characterization of linguistic variables we regard *F* as 'LARGE' in *X*. Using the operations of concentration and dilation

$F^{1/2}$ may be treated as 'More or less LARGE',
F^2 may be treated as 'Very LARGE',
F^4 may be treated as 'Very very LARGE'.

The proposed fuzzy mean difference divergence measures are used to calculate the degree of divergence between these fuzzy sets. The divergence values have been calculated by using Eqs. (4.1)–(4.18) between different fuzzy sets. The comparative results are summarized in Table 4.3. The following abbreviated notions are used in Table 4.3.

L.—LARGE
M.L.L.—More or less LARGE
V.L.—Very LARGE
V.V.L.—Very very LARGE

From the viewpoint of mathematical operations and the characterization of linguistic variables, the divergence between the above fuzzy sets has the following requirements:

$$\mathrm{D(L., M.L.L.) < D(L., V.L.) < D(L., V.V.L.)}, \tag{4.68}$$

$$\mathrm{D(M.L.L., L.) < D(M.L.L., V.L.) < D(M.L.L., V.V.L.)}, \tag{4.69}$$

$$\mathrm{D(V.L., V.V.L.) < D(V.L., L.) < D(V.L., M.L.L.)}, \tag{4.70}$$

$$\mathrm{D(V.V.L., V.L.) < D(V.V.L., L.) < D(V.V.L., M.L.L.)}. \tag{4.71}$$

It can be seen from the numerical results presented in Table 4.3 that the proposed fuzzy mean difference divergence measures for Eqs. (4.1)–(4.18) satisfy the requirement of Eqs. (4.68)–(4.71). Thus, the proposed fuzzy mean difference divergence measures are consistent in the application with compound linguistic measures.

Table 4.3 Divergence values calculated by Eqs. (4.1)–(4.18)

L.	M.L.L.	V.L.	V.V.L.	L.	M.L.L.	V.L.	V.V.L.	L.	M.L.L.	V.L.	V.V.L.	L.	M.L.L.	V.L.	V.V.L.
L.				M.L.L.				V.L.				V.V.L.			
$0.0000^{(6)}$	$0.0884^{(6)}$	$0.1304^{(6)}$	$0.4250^{(6)}$	$0.0884^{(6)}$	$0.0000^{(6)}$	$0.4523^{(6)}$	$0.8566^{(6)}$	$0.1304^{(6)}$	$0.4523^{(6)}$	$0.0000^{(6)}$	$0.1114^{(6)}$	$0.4250^{(6)}$	$0.8566^{(6)}$	$0.1114^{(6)}$	$0.0000^{(6)}$
$0.0000^{(7)}$	$0.2249^{(7)}$	$0.2945^{(7)}$	$0.9380^{(7)}$	$0.2249^{(7)}$	$0.0000^{(7)}$	$1.0043^{(7)}$	$1.9197^{(7)}$	$0.2945^{(7)}$	$1.0043^{(7)}$	$0.0000^{(7)}$	$0.2485^{(7)}$	$0.9380^{(7)}$	$1.9197^{(7)}$	$0.2485^{(7)}$	$0.0000^{(7)}$
$0.0000^{(8)}$	$0.3013^{(8)}$	$0.3836^{(8)}$	$1.2589^{(8)}$	$0.3013^{(8)}$	$0.0000^{(8)}$	$1.3277^{(8)}$	$2.5594^{(8)}$	$0.3836^{(8)}$	$1.3277^{(8)}$	$0.0000^{(8)}$	$0.3262^{(8)}$	$1.2589^{(8)}$	$2.5594^{(8)}$	$0.3262^{(8)}$	$0.0000^{(8)}$
$0.0000^{(9)}$	$0.1124^{(9)}$	$0.1666^{(9)}$	$0.5183^{(9)}$	$0.1124^{(9)}$	$0.0000^{(9)}$	$0.5617^{(9)}$	$0.6388^{(9)}$	$0.1666^{(9)}$	$0.5617^{(9)}$	$0.0000^{(9)}$	$0.1399^{(9)}$	$0.5183^{(9)}$	$0.6388^{(9)}$	$0.1399^{(9)}$	$0.0000^{(9)}$
$0.0000^{(10)}$	$0.1866^{(10)}$	$0.2499^{(10)}$	$0.7775^{(10)}$	$0.1866^{(10)}$	$0.0000^{(10)}$	$0.8426^{(10)}$	$1.5441^{(10)}$	$0.2499^{(10)}$	$0.8426^{(10)}$	$0.0000^{(10)}$	$0.2097^{(10)}$	$0.7775^{(10)}$	$1.5441^{(10)}$	$0.2097^{(10)}$	$0.0000^{(10)}$
$0.0000^{(11)}$	$0.4092^{(11)}$	$0.4998^{(11)}$	$1.5550^{(11)}$	$0.4092^{(11)}$	$0.0000^{(11)}$	$1.6852^{(11)}$	$3.0882^{(11)}$	$0.4998^{(11)}$	$1.6852^{(11)}$	$0.0000^{(11)}$	$0.4194^{(11)}$	$1.5550^{(11)}$	$3.0882^{(11)}$	$0.4194^{(11)}$	$0.0000^{(11)}$
$0.0000^{(12)}$	$0.0982^{(12)}$	$0.1195^{(12)}$	$0.3525^{(12)}$	$0.0982^{(12)}$	$0.0000^{(12)}$	$0.3903^{(12)}$	$0.6875^{(12)}$	$0.1195^{(12)}$	$0.3903^{(12)}$	$0.0000^{(12)}$	$0.0983^{(12)}$	$0.3525^{(12)}$	$0.6875^{(12)}$	$0.0983^{(12)}$	$0.0000^{(12)}$
$0.0000^{(13)}$	$0.1365^{(13)}$	$0.1641^{(13)}$	$0.5130^{(13)}$	$0.1365^{(13)}$	$0.0000^{(13)}$	$0.5520^{(13)}$	$1.0631^{(13)}$	$0.1641^{(13)}$	$0.5520^{(13)}$	$0.0000^{(13)}$	$0.1371^{(13)}$	$0.5130^{(13)}$	$1.0631^{(13)}$	$0.1371^{(13)}$	$0.0000^{(13)}$
$0.0000^{(14)}$	$0.2129^{(14)}$	$0.2532^{(14)}$	$0.8339^{(14)}$	$0.2129^{(14)}$	$0.0000^{(14)}$	$0.8754^{(14)}$	$1.7028^{(14)}$	$0.2532^{(14)}$	$0.8754^{(14)}$	$0.0000^{(14)}$	$0.2148^{(14)}$	$0.8339^{(14)}$	$1.7028^{(14)}$	$0.2148^{(14)}$	$0.0000^{(14)}$
$0.0000^{(15)}$	$0.3208^{(15)}$	$0.3694^{(15)}$	$1.1300^{(15)}$	$0.3208^{(15)}$	$0.0000^{(15)}$	$1.2329^{(15)}$	$2.2316^{(15)}$	$0.3694^{(15)}$	$1.2329^{(15)}$	$0.0000^{(15)}$	$0.3080^{(15)}$	$1.1300^{(15)}$	$2.2316^{(15)}$	$0.3080^{(15)}$	$0.0000^{(15)}$
$0.0000^{(16)}$	$0.0742^{(16)}$	$0.0833^{(16)}$	$0.2592^{(16)}$	$0.0742^{(16)}$	$0.0000^{(16)}$	$0.2809^{(16)}$	$0.9053^{(16)}$	$0.0833^{(16)}$	$0.2809^{(16)}$	$0.0000^{(16)}$	$0.0698^{(16)}$	$0.2592^{(16)}$	$0.9053^{(16)}$	$0.0698^{(16)}$	$0.0000^{(16)}$
$0.0000^{(17)}$	$0.1125^{(17)}$	$0.1279^{(17)}$	$0.4197^{(17)}$	$0.1125^{(17)}$	$0.0000^{(17)}$	$0.4426^{(17)}$	$1.2809^{(17)}$	$0.1279^{(17)}$	$0.4426^{(17)}$	$0.0000^{(17)}$	$0.1086^{(17)}$	$0.4197^{(17)}$	$1.2809^{(17)}$	$0.1086^{(17)}$	$0.0000^{(17)}$
$0.0000^{(18)}$	$0.1889^{(18)}$	$0.2170^{(18)}$	$0.7406^{(18)}$	$0.1889^{(18)}$	$0.0000^{(18)}$	$0.7660^{(18)}$	$1.9206^{(18)}$	$0.2170^{(18)}$	$0.7660^{(18)}$	$0.0000^{(18)}$	$0.1863^{(18)}$	$0.7406^{(18)}$	$1.9206^{(18)}$	$0.1863^{(18)}$	$0.0000^{(18)}$
$0.0000^{(19)}$	$0.2968^{(19)}$	$0.3332^{(19)}$	$1.0367^{(19)}$	$0.2968^{(19)}$	$0.0000^{(19)}$	$1.1235^{(19)}$	$2.4494^{(19)}$	$0.3332^{(19)}$	$1.1235^{(19)}$	$0.0000^{(19)}$	$0.2795^{(19)}$	$1.0367^{(19)}$	$2.4494^{(19)}$	$0.2795^{(19)}$	$0.0000^{(19)}$
$0.0000^{(20)}$	$0.0383^{(20)}$	$0.0446^{(20)}$	$0.1605^{(20)}$	$0.0383^{(20)}$	$0.0000^{(20)}$	$0.1617^{(20)}$	$0.3756^{(20)}$	$0.0446^{(20)}$	$0.1617^{(20)}$	$0.0000^{(20)}$	$0.0388^{(20)}$	$0.1605^{(20)}$	$0.3756^{(20)}$	$0.0388^{(20)}$	$0.0000^{(20)}$
$0.0000^{(21)}$	$0.1147^{(21)}$	$0.1337^{(21)}$	$0.4814^{(21)}$	$0.1147^{(21)}$	$0.0000^{(21)}$	$0.4851^{(21)}$	$1.0153^{(21)}$	$0.1337^{(21)}$	$0.4851^{(21)}$	$0.0000^{(21)}$	$0.1165^{(21)}$	$0.4814^{(21)}$	$1.0153^{(21)}$	$0.1165^{(21)}$	$0.0000^{(21)}$
$0.0000^{(22)}$	$0.2226^{(22)}$	$0.2499^{(22)}$	$0.7775^{(22)}$	$0.2226^{(22)}$	$0.0000^{(22)}$	$0.8426^{(22)}$	$1.5441^{(22)}$	$0.2499^{(22)}$	$0.8426^{(22)}$	$0.0000^{(22)}$	$0.2097^{(22)}$	$0.7775^{(22)}$	$1.5441^{(22)}$	$0.2097^{(22)}$	$0.0000^{(22)}$
$0.0000^{(23)}$	$0.0764^{(23)}$	$0.0891^{(23)}$	$0.3209^{(23)}$	$0.0764^{(23)}$	$0.0000^{(23)}$	$0.3234^{(23)}$	$0.6397^{(23)}$	$0.0891^{(23)}$	$0.3234^{(23)}$	$0.0000^{(23)}$	$0.0777^{(23)}$	$0.3209^{(23)}$	$0.6397^{(23)}$	$0.0777^{(23)}$	$0.0000^{(23)}$

For convenience we use the notation $*^{(i)}$ in Table 4.3 to present the divergence value computed from equation i.

4.6 Concluding Remarks

In this chapter, we have established a sequence of fuzzy mean difference divergence measures and inequalities among some of the proposed fuzzy mean difference divergence measures. An application of the proposed divergence measures is provided in the field of pattern recognition. The consistency of these divergence measures in application with compound linguistic variables is shown using a numerical example. Our results show that the fuzzy mean difference divergence measures are much simpler with the difference of the means involved.

References

1. Singh RP, Tomar VP (2014) On fuzzy mean divergence measures and their inequalities. In: 5th national conference at MIT, Academy of Engineering, Alandi(D), Pune, Maharashtra
2. Taneja IJ (2012) Inequalities having seven means and proportionality relations. http://arxiv.org/pdf/1203.2288v1.pdf. Accessed 18 April 2015
3. Taneja IJ (2005) Refinement of inequalities among means. http://arxiv.org/pdf/math/0505192v2.pdf. Accessed 20 April 2015
4. Bhandari D, Pal NR (1993) Some new information measures for fuzzy sets. Inf Sci 67 (3):209–228
5. Tomar VP, Ohlan A (2014a) Two new parametric generalized *R*—norm fuzzy information measures. Int J Comput Appl 93(13):22–27
6. Tomar VP, Ohlan A (2014b) Sequence of inequalities among fuzzy mean difference divergence measures and their applications. SpringerPlus 3 (623):1–20

Chapter 5
Applications of Generalized Fuzzy Divergence Measure in Multi-criteria Decision-Making and Pattern Recognition

Fuzzy divergence measures have broad applications in many areas such as pattern recognition, fuzzy clustering, decision-making, signal and image processing, speech recognition, bioinformatics, fuzzy aircraft control, feature selection, etc. This chapter provides a new parametric generalized measure of fuzzy divergence along with its applications in the context of pattern recognition and multi-criteria decision-making. In addition, the efficiency of the proposed measure is presented by proving its elegant properties.

In Sect. 5.1, we propose a new parametric measure of fuzzy divergence corresponding to Taneja [1] divergence measure. Following this, particular cases are discussed in the same section. Some more elegant properties of the proposed measure are studied in a number of theorems in Sect. 5.2. The applications of the proposed generalized divergence measure to pattern recognition and multi-criteria decision-making with numerical examples are then discussed in Sect. 5.3. Finally, the concluding remarks are drawn in Sect. 5.4. It is noted that the presence of the parameter in the proposed divergence measure provides greater flexibility for pattern recognition and multi-criteria decision-making.

5.1 A New Parametric Generalized Measure of Fuzzy Divergence

A number of improvements and generalizations came in the literature of fuzzy divergence measures [2] in the last decade. Taneja [1] proposed a parametric measure of divergence given by

A. Ohlan and R. Ohlan, *Generalizations of Fuzzy Information Measures*,
DOI 10.1007/978-3-319-45928-8_5

$$L_t(P,Q) = \sum_{i=1}^{n} \frac{(p_i - q_i)^2 (p_i + q_i)^t}{2^t (\sqrt{p_i q_i})^{t+1}}, \quad (P,Q) \in \Gamma_n \times \Gamma_n, t \in Z$$
$$\text{where } \Gamma_n = \left\{ P = (p_1, p_2, \ldots, p_n) \Big/ p_i > 0, \sum_{i=1}^{n} p_i = 1 \right\}, \quad n \geq 2. \tag{5.1}$$

We now propose a parametrically generalized measure of divergence between two fuzzy sets *A*, *B* of universe of discourse $X = \{x_1, x_2, x_3, \ldots, x_n\}$ having the membership values $\mu_A(x_i)$, $\mu_B(x_i) \in (0,1)$ corresponding to Eq. (5.1). It is given by

$$L_t(A,B) = \sum_{i=1}^{n} \frac{(\mu_A(x_i) - \mu_B(x_i))^2}{2^t} \left[\frac{(\mu_A(x_i) + \mu_B(x_i))^t}{\left(\sqrt{\mu_A(x_i)\mu_B(x_i)}\right)^{t+1}} + \frac{(2 - \mu_A(x_i) - \mu_B(x_i))^t}{\left(\sqrt{(1-\mu_A(x_i))(1-\mu_B(x_i))}\right)^{t+1}} \right]$$
$$\text{for } t = 0, 1, 2, \ldots \tag{5.2}$$

Theorem 5.1 *$L_t(A,B)$ is a valid measure of fuzzy divergence.*

Proof To prove the validity of $L_t(A,B)$ we use the definition of fuzzy divergence measure given by Couso et al. [3] as

If X is a universe of discourse and $F(X)$ is the set of all fuzzy subsets of X, a mapping $D : F(X) \times F(X) \to R$ is a fuzzy divergence measure if and only if for each $A, B, C \in F(X)$, the following axioms hold:

$$d_1 : D(A,B) = D(B,A)$$
$$d_2 : D(A,A) = 0$$
$$d_3 : \max\{D(A \cup C, B \cup C), D(A \cap C, B \cap C)\} \leq D(A,B)$$

Non-negativity of $D(A,B)$ is a natural assumption.

It is clear from Eq. (5.2) that

(i) $L_t(A,B) = L_t(B,A)$

(ii) $L_t(A,B) = 0$ if $\mu_A(x_i) = \mu_B(x_i)$

(iii) We now check the axiom d_3 of definition of Couso et al. [3].
We divide X into the following six subsets:

$$X_1 = \{x/x \in X, \ \mu_A(x) \leq \mu_B(x) \leq \mu_C(x)\},$$
$$X_2 = \{x/x \in X, \ \mu_A(x) \leq \mu_C(x) < \mu_B(x)\},$$
$$X_3 = \{x/x \in X, \ \mu_B(x) < \mu_A(x) \leq \mu_C(x)\},$$
$$X_4 = \{x/x \in X, \ \mu_B(x) \leq \mu_C(x) < \mu_A(x)\},$$
$$X_5 = \{x/x \in X, \ \mu_C(x) < \mu_A(x) \leq \mu_B(x)\},$$
$$X_6 = \{x/x \in X, \ \mu_C(x) < \mu_B(x) < \mu_A(x)\}.$$

In set X_1, $A \cup C$ =Union of A and $C \Leftrightarrow \mu_{A\cup C}(x) = \max\{\mu_A(x), \mu_C(x)\} = \mu_C(x)$;
$B \cup C$ = Union of B and $C \Leftrightarrow \mu_{B\cup C}(x) = \max\{\mu_B(x), \mu_C(x)\} = \mu_C(x)$;
$A \cap C$ = Intersection of A and $C \Leftrightarrow \mu_{A\cap C}(x) = \min\{\mu_A(x), \mu_C(x)\} = \mu_A(x)$;
$B \cap C$ = Intersection of B and $C \Leftrightarrow \mu_{B\cap C}(x) = \min\{\mu_B(x), \mu_C(x)\} = \mu_B(x)$;

$$L_t(A\cup C, B\cup C) = L_t(C, C) = 0,$$
$$L_t(A\cap C, B\cap C) = L_t(A, B),$$

So, $\max\{L_t(A\cup C, B\cup C), L_t(A\cap C, B\cap C)\} = L_t(A, B)$.
Similarly, in the sets X_2, X_3, X_4, X_5, X_6 we have $\max\{L_t(A\cup C, B\cup C), L_t(A\cap C, B\cap C)\} \leq L_t(A, B)$.
Thus, $\max\{L_t(A\cup C, B\cup C), L_t(A\cap C, B\cap C)\} \leq L_t(A, B)$ for all $A, B, C \in F(X)$.

(iv) We now show that $L_t(A, B) > 0$.

Consider two standard fuzzy sets $A = (0.3, 0.4, 0.2, 0.1, 0.5)$, $B = (0.2, 0.2, 0.3, 0.4, 0.4)$.

We take $L_t(A, B) = \sum_{i=1}^{n} e_i$, where $e_i = \frac{(\mu_A(x_i)-\mu_B(x_i))^2}{2^t}\left[\frac{(\mu_A(x_i)+\mu_B(x_i))^t}{\left(\sqrt{\mu_A(x_i)\mu_B(x_i)}\right)^{t+1}} + \frac{(2-\mu_A(x_i)-\mu_B(x_i))^t}{\left(\sqrt{(1-\mu_A(x_i))(1-\mu_B(x_i))}\right)^{t+1}}\right]$
for $i = 1, 2, 3, \ldots n$.

Table 5.1 gives positive values of $L_t(A, B)$ for any value of $t = 0, 1, 2, \ldots$. with two standard fuzzy sets considered above. It is clear from the results presented in Table 5.1 that for arbitrary values of $t \geq 0$, $L_t(A, B) > 0$.

Hence in view of the definition of fuzzy divergence measure of Couso et al. [3], $L_t(A, B)$ is a valid measure of fuzzy divergence.

In particular,

(i) For $t = -1$, $L_t(A, B)$ reduces to $2\Delta(A, B)$, where $\Delta(A, B)$ is the fuzzy triangular divergence measure of Singh and Tomar [4].
(ii) For $t = 1$, $L_t(A, B)$ reduces to $\frac{1}{2}\psi(A, B)$, where $\psi(A, B)$ is the fuzzy Chi-square divergence measure of Singh and Tomar [4].
(iii) For $t = 2$, $L_t(A, B)$ reduces to $\frac{1}{2}\psi_M(A, B)$, where $\psi_M(A, B)$ is the symmetric chi-square, arithmetic and geometric mean divergence measure of Kumar and Johnson [5] in fuzzy environment.

Table 5.1 Positive values of $L_t(A, B)$ for values of $t = 0, 1, 2\ldots$

t	e_1	e_2	e_3	e_4	e_5	$L_t(A, B)$
0	0.0542	0.1992	0.0542	0.5725	0.0406	0.9207
1	0.0551	0.2083	0.0551	0.7125	0.0408	1.0718
2	0.0177	0.2267	0.0177	0.1979	0.0410	0.5010
3	0.0568	0.0228	0.0568	1.0091	0.0413	1.1868
7	0.0484	0.2756	0.0484	0.1434	0.0425	0.5583
100	0.3310	51.2290	0.3310	2.2091e + 009	0.0693	2.2091e + 009

5.2 Properties of Proposed Generalized Fuzzy Divergence Measure

In this section we provide some more properties of the proposed generalized fuzzy divergence measure (5.2) in the following theorems. While proving these theorems we consider the separation of X into two parts X_1 and X_2 such that

$$X_1 = \{x/x \in X, \mu_A(x_i) \geq \mu_B(x_i)\} \tag{5.3}$$

$$\text{and } X_2 = \{x/x \in X, \mu_A(x_i) < \mu_B(x_i)\}. \tag{5.4}$$

Theorem 5.2

(a) $L_t(A \cup B, A \cap B) = L_t(A, B)$.
(b) $L_t(A \cup B, A) + L_t(A \cap B, A) = L_t(A, B)$.
(c) $L_t(A \cup B, C) \leq L_t(A, C) + L_t(B, C)$.

Proof

$$\begin{aligned}
5.2(\mathrm{a})\ L_t(A \cup B, A \cap B) &= \sum_{i=1}^{n} \frac{(\mu_{A\cup B}(x_i) - \mu_{A\cap B}(x_i))^2}{2^t} \left[\frac{(\mu_{A\cup B}(x_i) + \mu_{A\cap B}(x_i))^t}{\left(\sqrt{\mu_{A\cup B}(x_i)\mu_{A\cap B}(x_i)}\right)^{t+1}} + \frac{(2-\mu_{A\cup B}(x_i)-\mu_{A\cap B}(x_i))^t}{\left(\sqrt{(1-\mu_{A\cup B}(x_i))(1-\mu_{A\cap B}(x_i))}\right)^{t+1}} \right] \\
&= \sum_{X_1} \frac{(\mu_A(x_i) - \mu_B(x_i))^2}{2^t} \left[\frac{(\mu_A(x_i) + \mu_B(x_i))^t}{\left(\sqrt{\mu_A(x_i)\mu_B(x_i)}\right)^{t+1}} + \frac{(2 - \mu_A(x_i) - \mu_B(x_i))^t}{\left(\sqrt{(1-\mu_A(x_i))(1-\mu_B(x_i))}\right)^{t+1}} \right] \\
&\quad + \sum_{X_2} \frac{(\mu_B(x_i) - \mu_A(x_i))^2}{2^t} \left[\frac{(\mu_B(x_i) + \mu_A(x_i))^t}{\left(\sqrt{\mu_B(x_i)\mu_A(x_i)}\right)^{t+1}} + \frac{(2 - \mu_B(x_i) - \mu_A(x_i))^t}{\left(\sqrt{(1-\mu_B(x_i))(1-\mu_A(x_i))}\right)^{t+1}} \right] \\
&= \sum_{i=1}^{n} \frac{(\mu_A(x_i) - \mu_B(x_i))^2}{2^t} \left[\frac{(\mu_A(x_i) + \mu_B(x_i))^t}{\left(\sqrt{\mu_A(x_i)\mu_B(x_i)}\right)^{t+1}} + \frac{(2 - \mu_A(x_i) - \mu_B(x_i))^t}{\left(\sqrt{(1-\mu_A(x_i))(1-\mu_B(x_i))}\right)^{t+1}} \right] \\
&= L_t(A, B).
\end{aligned}$$

Hence, 5.2(a) holds.

$$\begin{aligned}
5.2(\mathrm{b})\ L_t(A \cup B, A) + L_t(A \cap B, A) &= \sum_{i=1}^{n} \frac{(\mu_{A\cup B}(x_i) - \mu_A(x_i))^2}{2^t} \left[\frac{(\mu_{A\cup B}(x_i) + \mu_A(x_i))^t}{\left(\sqrt{\mu_{A\cup B}(x_i)\mu_A(x_i)}\right)^{t+1}} + \frac{(2 - \mu_{A\cup B}(x_i) - \mu_A(x_i))^t}{\left(\sqrt{(1-\mu_{A\cup B}(x_i))(1-\mu_A(x_i))}\right)^{t+1}} \right] \\
&\quad + \sum_{i=1}^{n} \frac{(\mu_{A\cap B}(x_i) - \mu_A(x_i))^2}{2^t} \left[\frac{(\mu_{A\cap B}(x_i) + \mu_A(x_i))^t}{\left(\sqrt{\mu_{A\cap B}(x_i)\mu_A(x_i)}\right)^{t+1}} + \frac{(2 - \mu_{A\cap B}(x_i) - \mu_A(x_i))^t}{\left(\sqrt{(1-\mu_{A\cap B}(x_i))(1-\mu_A(x_i))}\right)^{t+1}} \right] \\
&= \sum_{X_2} \frac{(\mu_B(x_i) - \mu_A(x_i))^2}{2^t} \left[\frac{(\mu_B(x_i) + \mu_A(x_i))^t}{\left(\sqrt{\mu_B(x_i)\mu_A(x_i)}\right)^{t+1}} + \frac{(2 - \mu_B(x_i) - \mu_A(x_i))^t}{\left(\sqrt{(1-\mu_B(x_i))(1-\mu_A(x_i))}\right)^{t+1}} \right] \\
&\quad + \sum_{X_1} \frac{(\mu_B(x_i) - \mu_A(x_i))^2}{2^t} \left[\frac{(\mu_B(x_i) + \mu_A(x_i))^t}{\left(\sqrt{\mu_B(x_i)\mu_A(x_i)}\right)^{t+1}} + \frac{(2 - \mu_B(x_i) - \mu_A(x_i))^t}{\left(\sqrt{(1-\mu_B(x_i))(1-\mu_A(x_i))}\right)^{t+1}} \right] \\
&= L_t(A, B).
\end{aligned}$$

Hence, 5.2(b) holds.

5.2(c) $$L_t(A,C)+L_t(B,C)-L_t(A\cup B,C)=\sum_{i=1}^{n}\frac{(\mu_A(x_i)-\mu_C(x_i))^2}{2^t}\left[\frac{(\mu_A(x_i)+\mu_C(x_i))^t}{\left(\sqrt{\mu_A(x_i)\mu_C(x_i)}\right)^{t+1}}+\frac{(2-\mu_A(x_i)-\mu_C(x_i))^t}{\left(\sqrt{(1-\mu_A(x_i))(1-\mu_C(x_i))}\right)^{t+1}}\right]$$
$$+\sum_{i=1}^{n}\frac{(\mu_B(x_i)-\mu_C(x_i))^2}{2^t}\left[\frac{(\mu_B(x_i)+\mu_C(x_i))^t}{\left(\sqrt{\mu_B(x_i)\mu_C(x_i)}\right)^{t+1}}+\frac{(2-\mu_B(x_i)-\mu_C(x_i))^t}{\left(\sqrt{(1-\mu_B(x_i))(1-\mu_C(x_i))}\right)^{t+1}}\right]$$
$$-\sum_{i=1}^{n}\frac{(\mu_{A\cup B}(x_i)-\mu_C(x_i))^2}{2^t}\left[\frac{(\mu_{A\cup B}(x_i)+\mu_C(x_i))^t}{\left(\sqrt{\mu_{A\cup B}(x_i)\mu_C(x_i)}\right)^{t+1}}+\frac{(2-\mu_{A\cup B}(x_i)-\mu_C(x_i))^t}{\left(\sqrt{(1-\mu_{A\cup B}(x_i))(1-\mu_C(x_i))}\right)^{t+1}}\right]$$
$$=\sum_{X_2}\frac{(\mu_A(x_i)-\mu_C(x_i))^2}{2^t}\left[\frac{(\mu_A(x_i)+\mu_C(x_i))^t}{\left(\sqrt{\mu_A(x_i)\mu_C(x_i)}\right)^{t+1}}+\frac{(2-\mu_A(x_i)-\mu_C(x_i))^t}{\left(\sqrt{(1-\mu_A(x_i))(1-\mu_C(x_i))}\right)^{t+1}}\right]$$
$$+\sum_{X_1}\frac{(\mu_B(x_i)-\mu_C(x_i))^2}{2^t}\left[\frac{(\mu_B(x_i)+\mu_C(x_i))^t}{\left(\sqrt{\mu_B(x_i)\mu_C(x_i)}\right)^{t+1}}+\frac{(2-\mu_B(x_i)-\mu_C(x_i))^t}{\left(\sqrt{(1-\mu_B(x_i))(1-\mu_C(x_i))}\right)^{t+1}}\right]$$
$$\Rightarrow L_t(A,C)+L_t(B,C)\geq L_t(A\cup B,C)\geq 0.$$

Hence, 5.2(c) holds.

Theorem 5.3

(a) $L_t(A\cup B,C)+L_t(A\cap B,C)=L_t(A,C)+L_t(B,C)$.
(b) $L_t(A,A\cap B)=L_t(B,A\cup B)$.
(c) $L_t(A,A\cup B)=L_t(B,A\cap B)$.

Proof

5.3(a) $$L_t(A\cup B,C)+L_t(A\cap B,C)=\sum_{i=1}^{n}\frac{(\mu_{A\cup B}(x_i)-\mu_C(x_i))^2}{2^t}\left[\frac{(\mu_{A\cup B}(x_i)+\mu_C(x_i))^t}{\left(\sqrt{\mu_{A\cup B}(x_i)\mu_C(x_i)}\right)^{t+1}}+\frac{(2-\mu_{A\cup B}(x_i)-\mu_C(x_i))^t}{\left(\sqrt{(1-\mu_{A\cup B}(x_i))(1-\mu_C(x_i))}\right)^{t+1}}\right]$$
$$+\sum_{i=1}^{n}\frac{(\mu_{A\cap B}(x_i)-\mu_C(x_i))^2}{2^t}\left[\frac{(\mu_{A\cap B}(x_i)+\mu_C(x_i))^t}{\left(\sqrt{\mu_{A\cap B}(x_i)\mu_C(x_i)}\right)^{t+1}}+\frac{(2-\mu_{A\cap B}(x_i)-\mu_C(x_i))^t}{\left(\sqrt{(1-\mu_{A\cap B}(x_i))(1-\mu_C(x_i))}\right)^{t+1}}\right]$$
$$=\sum_{X_1}\frac{(\mu_A(x_i)-\mu_C(x_i))^2}{2^t}\left[\frac{(\mu_A(x_i)+\mu_C(x_i))^t}{\left(\sqrt{\mu_A(x_i)\mu_C(x_i)}\right)^{t+1}}+\frac{(2-\mu_A(x_i)-\mu_C(x_i))^t}{\left(\sqrt{(1-\mu_A(x_i))(1-\mu_C(x_i))}\right)^{t+1}}\right]$$
$$+\sum_{X_2}\frac{(\mu_B(x_i)-\mu_C(x_i))^2}{2^t}\left[\frac{(\mu_B(x_i)+\mu_C(x_i))^t}{\left(\sqrt{\mu_B(x_i)\mu_C(x_i)}\right)^{t+1}}+\frac{(2-\mu_B(x_i)-\mu_C(x_i))^t}{\left(\sqrt{(1-\mu_B(x_i))(1-\mu_C(x_i))}\right)^{t+1}}\right]$$
$$+\sum_{X_1}\frac{(\mu_B(x_i)-\mu_C(x_i))^2}{2^t}\left[\frac{(\mu_B(x_i)+\mu_C(x_i))^t}{\left(\sqrt{\mu_B(x_i)\mu_C(x_i)}\right)^{t+1}}+\frac{(2-\mu_B(x_i)-\mu_C(x_i))^t}{\left(\sqrt{(1-\mu_B(x_i))(1-\mu_C(x_i))}\right)^{t+1}}\right]$$
$$+\sum_{X_2}\frac{(\mu_A(x_i)-\mu_C(x_i))^2}{2^t}\left[\frac{(\mu_A(x_i)+\mu_C(x_i))^t}{\left(\sqrt{\mu_A(x_i)\mu_C(x_i)}\right)^{t+1}}+\frac{(2-\mu_A(x_i)-\mu_C(x_i))^t}{\left(\sqrt{(1-\mu_A(x_i))(1-\mu_C(x_i))}\right)^{t+1}}\right]$$
$$=\sum_{i=1}^{n}\frac{(\mu_A(x_i)-\mu_C(x_i))^2}{2^t}\left[\frac{(\mu_A(x_i)+\mu_C(x_i))^t}{\left(\sqrt{\mu_A(x_i)\mu_C(x_i)}\right)^{t+1}}+\frac{(2-\mu_A(x_i)-\mu_C(x_i))^t}{\left(\sqrt{(1-\mu_A(x_i))(1-\mu_C(x_i))}\right)^{t+1}}\right]$$
$$+\sum_{i=1}^{n}\frac{(\mu_B(x_i)-\mu_C(x_i))^2}{2^t}\left[\frac{(\mu_B(x_i)+\mu_C(x_i))^t}{\left(\sqrt{\mu_B(x_i)\mu_C(x_i)}\right)^{t+1}}+\frac{(2-\mu_B(x_i)-\mu_C(x_i))^t}{\left(\sqrt{(1-\mu_B(x_i))(1-\mu_C(x_i))}\right)^{t+1}}\right]$$
$$=L_t(A,C)+L_t(B,C).$$

Hence, 5.3(a) holds.

5.3(b)
$$L_t(A, A\cap B) = \sum_{i=1}^{n} \frac{(\mu_A(x_i) - \mu_{A\cap B}(x_i))^2}{2^t} \left[\frac{(\mu_A(x_i) + \mu_{A\cap B}(x_i))^t}{\left(\sqrt{\mu_A(x_i)\mu_{A\cap B}(x_i)}\right)^{t+1}} + \frac{(2-\mu_A(x_i)-\mu_{A\cap B}(x_i))^t}{\left(\sqrt{(1-\mu_A(x_i))(1-\mu_{A\cap B}(x_i))}\right)^{t+1}} \right]$$
$$= \sum_{X_1} \frac{(\mu_A(x_i) - \mu_B(x_i))^2}{2^t} \left[\frac{(\mu_A(x_i) + \mu_B(x_i))^t}{\left(\sqrt{\mu_A(x_i)\mu_B(x_i)}\right)^{t+1}} + \frac{(2-\mu_A(x_i)-\mu_B(x_i))^t}{\left(\sqrt{(1-\mu_A(x_i))(1-\mu_B(x_i))}\right)^{t+1}} \right]$$

Now

$$L_t(B, A\cup B) = \sum_{i=1}^{n} \frac{(\mu_B(x_i) - \mu_{A\cup B}(x_i))^2}{2^t} \left[\frac{(\mu_B(x_i) + \mu_{A\cup B}(x_i))^t}{\left(\sqrt{\mu_B(x_i)\mu_{A\cup B}(x_i)}\right)^{t+1}} + \frac{(2-\mu_B(x_i)-\mu_{A\cup B}(x_i))^t}{\left(\sqrt{(1-\mu_B(x_i))(1-\mu_{A\cup B}(x_i))}\right)^{t+1}} \right]$$
$$= \sum_{X_1} \frac{(\mu_B(x_i) - \mu_A(x_i))^2}{2^t} \left[\frac{(\mu_B(x_i) + \mu_A(x_i))^t}{\left(\sqrt{\mu_B(x_i)\mu_A(x_i)}\right)^{t+1}} + \frac{(2-\mu_B(x_i)-\mu_A(x_i))^t}{\left(\sqrt{(1-\mu_B(x_i))(1-\mu_A(x_i))}\right)^{t+1}} \right]$$
$$= L_t(A, A\cap B).$$

Hence, 5.3(b) holds.

5.3(c)
$$L_t(A, A\cup B) = \sum_{i=1}^{n} \frac{(\mu_A(x_i) - \mu_{A\cup B}(x_i))^2}{2^t} \left[\frac{(\mu_A(x_i) + \mu_{A\cup B}(x_i))^t}{\left(\sqrt{\mu_A(x_i)\mu_{A\cup B}(x_i)}\right)^{t+1}} + \frac{(2-\mu_A(x_i)-\mu_{A\cup B}(x_i))^t}{\left(\sqrt{(1-\mu_A(x_i))(1-\mu_{A\cup B}(x_i))}\right)^{t+1}} \right]$$
$$= \sum_{X_2} \frac{(\mu_A(x_i) - \mu_B(x_i))^2}{2^t} \left[\frac{(\mu_A(x_i) + \mu_B(x_i))^t}{\left(\sqrt{\mu_A(x_i)\mu_B(x_i)}\right)^{t+1}} + \frac{(2-\mu_A(x_i)-\mu_B(x_i))^t}{\left(\sqrt{(1-\mu_A(x_i))(1-\mu_B(x_i))}\right)^{t+1}} \right]$$

Now

$$L_t(B, A\cap B) = \sum_{i=1}^{n} \frac{(\mu_B(x_i) - \mu_{A\cap B}(x_i))^2}{2^t} \left[\frac{(\mu_B(x_i) + \mu_{A\cap B}(x_i))^t}{\left(\sqrt{\mu_B(x_i)\mu_{A\cap B}(x_i)}\right)^{t+1}} + \frac{(2-\mu_B(x_i)-\mu_{A\cap B}(x_i))^t}{\left(\sqrt{(1-\mu_B(x_i))(1-\mu_{A\cap B}(x_i))}\right)^{t+1}} \right]$$
$$= \sum_{X_2} \frac{(\mu_B(x_i) - \mu_A(x_i))^2}{2^t} \left[\frac{(\mu_B(x_i) + \mu_A(x_i))^t}{\left(\sqrt{\mu_B(x_i)\mu_A(x_i)}\right)^{t+1}} + \frac{(2-\mu_B(x_i)-\mu_A(x_i))^t}{\left(\sqrt{(1-\mu_B(x_i))(1-\mu_A(x_i))}\right)^{t+1}} \right]$$
$$= L_t(A, A\cup B).$$

Hence, 5.3(c) holds.

Theorem 5.4

(a) $L_t(A,\overline{A}) = L_t(\overline{A},A)$.
(b) $L_t(\overline{A},\overline{B}) = L_t(A,B)$.
(c) $L_t(A,\overline{B}) = L_t(\overline{A},B)$.
(d) $L_t(A,B) + L_t(\overline{A},B) = L_t(\overline{A},\overline{B}) + L_t(A,\overline{B})$.

Proof

$$5.4(a)\ L_t(A,\overline{A}) = \sum_{i=1}^{n} \frac{(\mu_A(x_i) - \mu_{\overline{A}}(x_i))^2}{2^t}\left[\frac{(\mu_A(x_i)+\mu_{\overline{A}}(x_i))^t}{\left(\sqrt{\mu_A(x_i)\mu_{\overline{A}}(x_i)}\right)^{t+1}} + \frac{(2-\mu_A(x_i)-\mu_{\overline{A}}(x_i))^t}{\left(\sqrt{(1-\mu_A(x_i))(1-\mu_{\overline{A}}(x_i))}\right)^{t+1}}\right]$$
$$= \sum_{i=1}^{n} \frac{(\mu_A(x_i) - 1 + \mu_A(x_i))^2}{2^t}\left[\frac{(\mu_A(x_i)+1-\mu_A(x_i))^t}{\left(\sqrt{\mu_A(x_i)(1-\mu_A(x_i))}\right)^{t+1}} + \frac{(2-\mu_A(x_i)-1+\mu_A(x_i))^t}{\left(\sqrt{(1-\mu_A(x_i))\mu_A(x_i)}\right)^{t+1}}\right]$$
$$= \sum_{i=1}^{n} \frac{(2\mu_A(x_i)-1)^2}{2^{t-1}\left(\sqrt{\mu_A(x_i)(1-\mu_A(x_i))}\right)^{t+1}}$$

Now

$$L_t(\overline{A},A) = \sum_{i=1}^{n} \frac{(\mu_{\overline{A}}(x_i) - \mu_A(x_i))^2}{2^t}\left[\frac{(\mu_{\overline{A}}(x_i)+\mu_A(x_i))^t}{\left(\sqrt{\mu_{\overline{A}}(x_i)\mu_A(x_i)}\right)^{t+1}} + \frac{(2-\mu_{\overline{A}}(x_i)-\mu_A(x_i))^t}{\left(\sqrt{(1-\mu_{\overline{A}}(x_i))(1-\mu_A(x_i))}\right)^{t+1}}\right]$$
$$= \sum_{i=1}^{n} \frac{(1 - \mu_A(x_i) + \mu_A(x_i))^2}{2^t}\left[\frac{(1-\mu_A(x_i)+\mu_A(x_i))^t}{\left(\sqrt{(1-\mu_A(x_i))\mu_A(x_i)}\right)^{t+1}} + \frac{(2-1+\mu_A(x_i)-\mu_A(x_i))^t}{\left(\sqrt{\mu_A(x_i)(1-\mu_A(x_i))}\right)^{t+1}}\right]$$
$$= \sum_{i=1}^{n} \frac{(2\mu_A(x_i)-1)^2}{2^{t-1}\left(\sqrt{\mu_A(x_i)(1-\mu_A(x_i))}\right)^{t+1}}$$
$$= L_t(A,\overline{A}).$$

Hence, 5.4(a) holds.

$$5.4(b)\ L_t(\overline{A},\overline{B}) = \sum_{i=1}^{n} \frac{(\mu_{\overline{A}}(x_i) - \mu_{\overline{B}}(x_i))^2}{2^t}\left[\frac{(\mu_{\overline{A}}(x_i)+\mu_{\overline{B}}(x_i))^t}{\left(\sqrt{\mu_{\overline{A}}(x_i)\mu_{\overline{B}}(x_i)}\right)^{t+1}} + \frac{(2-\mu_{\overline{A}}(x_i)-\mu_{\overline{B}}(x_i))^t}{\left(\sqrt{(1-\mu_{\overline{A}}(x_i))(1-\mu_{\overline{B}}(x_i))}\right)^{t+1}}\right]$$
$$= \sum_{i=1}^{n} \frac{(\mu_B(x_i) - \mu_A(x_i))^2}{2^t}\left[\frac{(2-\mu_A(x_i)-\mu_B(x_i))^t}{\left(\sqrt{(1-\mu_A(x_i))(1-\mu_B(x_i))}\right)^{t+1}} + \frac{(\mu_A(x_i)+\mu_B(x_i))^t}{\left(\sqrt{\mu_A(x_i)\mu_B(x_i)}\right)^{t+1}}\right]$$
$$= L_t(A,B).$$

Hence, 5.4(b) holds.

$$5.4(\mathrm{c})\ L_t(A,\overline{B}) = \sum_{i=1}^{n} \frac{(\mu_A(x_i) - \mu_{\overline{B}}(x_i))^2}{2^t}\left[\frac{(\mu_A(x_i) + \mu_{\overline{B}}(x_i))^t}{\left(\sqrt{\mu_A(x_i)\mu_{\overline{B}}(x_i)}\right)^{t+1}} + \frac{(2 - \mu_A(x_i) - \mu_{\overline{B}}(x_i))^t}{\left(\sqrt{(1-\mu_A(x_i))(1-\mu_{\overline{B}}(x_i))}\right)^{t+1}}\right]$$

$$= \sum_{i=1}^{n} \frac{(\mu_A(x_i) + \mu_B(x_i) - 1)^2}{2^t}\left[\frac{(1 + \mu_A(x_i) - \mu_B(x_i))^t}{\left(\sqrt{\mu_A(x_i)(1-\mu_B(x_i))}\right)^{t+1}} + \frac{(1 - \mu_A(x_i) + \mu_B(x_i))^t}{\left(\sqrt{(1-\mu_A(x_i))\mu_B(x_i)}\right)^{t+1}}\right]$$

Now

$$L_t(\overline{A},B) = \sum_{i=1}^{n} \frac{(\mu_{\overline{A}}(x_i) - \mu_B(x_i))^2}{2^t}\left[\frac{(\mu_{\overline{A}}(x_i) + \mu_B(x_i))^t}{\left(\sqrt{\mu_{\overline{A}}(x_i)\mu_B(x_i)}\right)^{t+1}} + \frac{(2 - \mu_{\overline{A}}(x_i) - \mu_B(x_i))^t}{\left(\sqrt{(1-\mu_{\overline{A}}(x_i))(1-\mu_B(x_i))}\right)^{t+1}}\right]$$

$$= \sum_{i=1}^{n} \frac{(1 - \mu_A(x_i) - \mu_B(x_i))^2}{2^t}\left[\frac{(1 - \mu_A(x_i) + \mu_B(x_i))^t}{\left(\sqrt{(1-\mu_A(x_i))\mu_B(x_i)}\right)^{t+1}} + \frac{(1 + \mu_A(x_i) - \mu_B(x_i))^t}{\left(\sqrt{\mu_A(x_i)(1-\mu_B(x_i))}\right)^{t+1}}\right]$$

$$= L_t(A,\overline{B}).$$

Hence, 5.4(c) holds.

5.4(d) It obviously follows from 5.4(b) and 5.4(c).

5.3 Applications of the Proposed Generalized Fuzzy Divergence Measure

In recent years, the applications of the fuzzy divergence measure have been given in different areas: Poletti et al. [6] in bioinformatics; Bhandari et al. [7], Fan et al. [8] and Bhatia and Singh [9] in image thresholding; Ghosh et al. [10] in automated leukocyte recognition. We present the application of the proposed generalized fuzzy divergence measure in the context of pattern recognition and multi-criteria decision-making.

5.3.1 *Pattern Recognition*

Dealing with uncertainties is a common problem in pattern recognition and the use of fuzzy set theory has given rise to a lot of new methods of pattern recognition. We now present application of the proposed generalized fuzzy divergence measure in the context of pattern recognition.

In order to demonstrate the applications of the introduced generalized fuzzy divergence measure in pattern recognition, suppose that we are given m known patterns $P_1, P_2, P_3, \ldots, P_m$ which have classifications $C_1, C_2, C_3, \ldots, C_m$ respectively. The patterns are represented by the following fuzzy sets in the universe of discourse $X = \{x_1, x_2, x_3, \ldots, x_n\}$:

$$P_i = \{\langle x_j, \mu_{P_i}(x_j)\rangle / x_j \in X\}$$

where $i = 1, 2, \ldots, m$ and $j = 1, 2, \ldots, n$.

Given an unknown pattern Q, represented by the fuzzy set

$$Q_i = \{\langle x_j, \mu_{Q_i}(x_j)\rangle / x_j \in X\}.$$

Our aim here is to classify Q to one of the classes $C_1, C_2, C_3, \ldots, C_m$. According to the principle of minimum divergence/discrimination information between fuzzy sets, the process of assigning Q to C_{k^*} is described by

$$k^* = \arg\min_k \{D(P_k, Q)\}.$$

According to this algorithm, the given pattern can be recognized so that the best class can be selected. It is a practical application of minimum divergence measure principle of Shore and Gray [11] to pattern recognition.

A Numerical Example

We consider the problem having four known patterns P_1, P_2, P_3 and P_4 which have classifications C_1, C_2, C_3 and C_4 respectively. These are represented by the following fuzzy sets in the universe of discourse $X = \{x_1, x_2, x_3\}$:

$$P_1 = \{\langle x_1, 0.7\rangle, \langle x_2, 0.3\rangle, \langle x_3, 0.1\rangle\},$$
$$P_2 = \{\langle x_1, 0.4\rangle, \langle x_2, 0.2\rangle, \langle x_3, 0.5\rangle\},$$
$$P_3 = \{\langle x_1, 0.6\rangle, \langle x_2, 0.3\rangle, \langle x_3, 0.8\rangle\},$$
$$P_4 = \{\langle x_1, 0.7\rangle, \langle x_2, 0.5\rangle, \langle x_3, 0.8\rangle\}$$

Given an unknown pattern Q, represented by the fuzzy set

$$Q = \{\langle x_1, 0.5\rangle, \langle x_2, 0.4\rangle, \langle x_3, 0.9\rangle\},$$

Our aim now is to classify Q to one of the classes C_1, C_2, C_3 and C_4. From the formula (5.2), we can compute the values of generalized fuzzy divergence measure $L_t(P_k, Q), k = \{1, 2, 3, 4\}$, for any $t \geq 0$ and are presented in Table 5.2 as follows.

It is observed that Q can be classified to C_3 correctly.

Table 5.2 Computed values of generalized fuzzy divergence measure $L_t(P_k, Q)$, $k = \{1, 2, 3, 4\}$ for any $t \geq 0$

	$t = 0$	$t = 5$	$t = 10$	$t = 20$	$t = 100$
$L_t(P_1, Q)$	4.4819	55.1097	705.8971	1.1670e + 005	6.5308e + 022
$L_t(P_2, Q)$	1.1939	3.6978	14.2475	256.6162	4.1520e + 012
$L_t(P_3, Q)$	**0.1674**	**0.1946**	**0.2301**	**0.3389**	**25.7186**
$L_t(P_4, Q)$	0.2940	0.3424	0.4027	0.5735	28.4963

5.3.2 Multi-criteria Decision-Making

In many situations decision-makers have imprecise information about alternatives with respect to criterion. Decision-making deals with the problem of choosing the best alternative, that is, the one with the highest degree of satisfaction for all the appropriate criteria or goals. Multi-criteria decision-making (MCDM) is a well-established branch of decision-making that permits decision-makers to rank and select alternatives according to different criteria. MCDM techniques are important and popular mathematical methods used in a variety of human activities. We now turn to discuss the applications of the proposed generalized fuzzy divergence in the context of multi-criteria decision-making. For this we provide a method to solve multi-criteria decision-making problems with the help of the proposed generalized fuzzy divergence measure.

Let $M = \{M_1, M_2, \ldots, M_m\}$ be a set of options, $C = \{C_1, C_2, \ldots, C_n\}$ be the set of criteria. The characteristics of the option M_i in terms of criteria C are represented by the following FSs:

$$M_i = \{\langle C_j, \mu_{ij}\rangle, C_j \in C\}, \quad i = 1, 2, 3, \ldots, m \quad \text{and} \quad j = 1, 2, 3, \ldots, n$$

where μ_{ij} indicates the degree that the option M_i satisfies the criterion C_j.

We introduce the approach to solve the above multi-criteria fuzzy decision-making problem using Eq. (5.2). The computational procedure of the same is as follows.

Step 1 Find out the positive-ideal solution M^+ and negative-ideal solution M^-: $M^+ = \{\langle\mu_1^+\rangle, \langle\mu_2^+\rangle, \ldots, \langle\mu_n^+\rangle\}$ and $M^- = \{\langle\mu_1^-\rangle, \langle\mu_2^-\rangle, \ldots, \langle\mu_n^-\rangle\}$, where for each $j = 1, 2, 3, \ldots, n$,

$$\left.\begin{aligned} \left\langle\mu_j^+\right\rangle &= \left\langle \max_i \mu_{ij}\right\rangle \\ \left\langle\mu_j^-\right\rangle &= \left\langle \min_i \mu_{ij}\right\rangle \end{aligned}\right\}. \tag{5.5}$$

Step 2 Calculate $L_t(M^+, M_i)$ and $L_t(M^-, M_i)$ using Eq. (5.2).

Step 3 Calculate the relative fuzzy divergence measure $L_t(M_i)$ of alternative M_i with respect to M^+ and M^-,

$$\text{where } L_t(M_i) = \frac{L_t(M^+, M_i)}{L_t(M^+, M_i) + L_t(M^-, M_i)}, \quad i = 1, 2, 3, \ldots, m. \tag{5.6}$$

Step 4 Rank the preference order of all alternatives according to the relative fuzzy divergence measure.

Step 5 Select the best alternative M_k with the smallest $L_t(M_k)$.

A Numerical Example

We now demonstrate the applicability of the proposed generalized fuzzy divergence measure to solve a real problem related to multi-criteria decision-making. For this we consider a customer decision-making problem in purchasing a car.

Example Consider a customer who wants to buy a car. Let five types of cars, i.e. the alternatives $M = \{M_1, M_2, M_3, M_4, M_5\}$ be available in the market. To buy a car the customer takes the following four criteria to decide: (i) Quality of Product (C_1), (ii) Price (C_2), (iii) Technical Capability (C_3) and (iv) Fuel Economy (C_4).

The five possible options are to be evaluated by the decision-maker under the above four criteria in the following form:

$$\begin{aligned}
M_1 &= \{\langle C_1, 0.5\rangle, \langle C_2, 0.3\rangle, \langle C_3, 0.4\rangle, \langle C_4, 0.7\rangle\},\\
M_2 &= \{\langle C_1, 0.2\rangle, \langle C_2, 0.7\rangle, \langle C_3, 0.6\rangle, \langle C_4, 0.6\rangle\},\\
M_3 &= \{\langle C_1, 0.8\rangle, \langle C_2, 0.5\rangle, \langle C_3, 0.9\rangle, \langle C_4, 0.2\rangle\},\\
M_4 &= \{\langle C_1, 0.6\rangle, \langle C_2, 0.4\rangle, \langle C_3, 0.7\rangle, \langle C_4, 0.5\rangle\},\\
M_5 &= \{\langle C_1, 0.6\rangle, \langle C_2, 0.5\rangle, \langle C_3, 0.5\rangle, \langle C_4, 0.7\rangle\}.
\end{aligned}$$

The stepwise computational procedure to solve the above multi-criteria fuzzy decision-making problem now goes as follows.

Step 1 The positive-ideal solution M^+ and negative-ideal solution M^- respectively are

$$\begin{aligned}
M^+ &= \{\langle C_1, 0.8\rangle, \langle C_2, 0.7\rangle, \langle C_3, 0.9\rangle, \langle C_4, 0.7\rangle\}\\
M^- &= \{\langle C_1, 0.2\rangle, \langle C_2, 0.3\rangle, \langle C_3, 0.4\rangle, \langle C_4, 0.2\rangle\}.
\end{aligned}$$

Step 2 Tables 5.3 and 5.4 show the calculated numerical values of $L_t(M^+, M_i)$ and $L_t(M^-, M_i)$ using Eq. (5.2) for $t \geq 0$.

Step 3 Calculated values of the relative fuzzy divergence measure $L_t(M_i)$ for $i = 1, 2, 3, 4, 5$ with $t \geq 0$ are presented in Table 5.5.

Step 4 According to the calculated numerical values of relative divergence for different values of $t \geq 0$, ranking order of alternatives is as follows:

Table 5.3 Calculated numerical values of $L_t(M^+, M_i)$, $t \geq 0$

	$t = 0$	$t = 1$	$t = 2$	$t = 10$	$t = 50$
$L_t(M^+, M_1)$	2.5625	3.1328	3.9030	39.7763	5.7344e + 007
$L_t(M^+, M_2)$	2.4168	2.9821	3.6882	21.1529	1.5765e + 005
$L_t(M^+, M_3)$	1.3494	1.5517	1.7893	6.0740	6.9636e + 003
$L_t(M^+, M_4)$	1.0336	1.1028	1.1792	2.1997	315.5988
$L_t(M^+, M_5)$	1.3242	1.5924	1.9454	14.4252	1.7236e + 006

Table 5.4 Calculated numerical values of $L_t(M^-, M_i)$, $t \geq 0$

	$t = 0$	$t = 1$	$t = 2$	$t = 10$	$t = 50$
$L_t(M^-, M_1)$	1.6054	1.8377	2.1085	6.8257	7.0090e + 003
$L_t(M^-, M_2)$	1.6063	1.7619	1.9354	4.8261	674.1698
$L_t(M^-, M_3)$	3.4082	4.3349	5.5650	54.1187	5.7470e + 007
$L_t(M^-, M_4)$	1.5981	1.7410	1.9003	4.1088	670.4259
$L_t(M^-, M_5)$	2.1347	2.4258	2.7643	8.5729	7.5828e + 003

Table 5.5 Computed values of relative divergence measure $L_t(M_i)$ for $i = 1, 2, 3, 4, 5$

	$t = 0$	$t = 1$	$t = 2$	$t = 10$	$t = 50$
$L_t(M_1)$	0.6148	0.6303	0.6493	0.8535	0.9999
$L_t(M_2)$	0.6007	0.6286	0.6558	0.8142	0.9957
$L_t(M_3)$	**0.2836**	**0.2636**	**0.2433**	**0.1009**	**1.2115e–004**
$L_t(M_4)$	0.3927	0.3878	0.3829	0.3487	0.3201
$L_t(M_5)$	0.3828	0.3963	0.4131	0.6272	0.9956

$$\text{For } t = 0,\ M_3 > M_5 > M_4 > M_2 > M_1.$$
$$\text{For } t = 1,\ M_3 > M_4 > M_5 > M_2 > M_1.$$
$$\text{For } t = 2,\ M_3 > M_4 > M_5 > M_1 > M_2.$$
$$\text{For } t = 10,\ M_3 > M_4 > M_5 > M_1 > M_2.$$
$$\text{For } t = 50,\ M_3 > M_4 > M_5 > M_2 > M_1.$$

Thus we here find that changes in values of t carry an alteration in ranking, but leave the best alternative unchanged. So M_3 is the most preferable alternative.

5.4 Concluding Remarks

This chapter proposed and validated a new parametric generalized measure of fuzzy divergence. The particular cases have been discussed in detail along with some of the properties of this fuzzy divergence measure. The applications of the proposed generalized fuzzy divergence measure are shown to pattern recognition and multi-criteria fuzzy decision-making. Finally, a numerical example is provided to illustrate the multi-criteria decision-making process proposed by us. We note that the presence of the parameter in the proposed divergence measure provides greater flexibility for pattern recognition and multi-criteria decision-making.

References

1. Taneja IJ (2013) Seven means, generalized triangular discrimination and generating divergence measures. Information 4:198–239
2. Ohlan A (2015) A new generalized fuzzy divergence measure and applications. Fuzzy Inf Eng 7(4):507–523
3. Couso I, Janis V, Montes S (2000) Fuzzy divergence measures. Acta Univ M Belii 8:21–26
4. Singh RP, Tomar VP (2008) On fuzzy divergence measures and their inequalities. In: Proceedings of 10th national conference of ISITA, pp 41–43
5. Kumar P, Johnson A (2005) On a symmetric divergence measure and information inequalities. J Inequalities Pure Appl Math 6(3):1–13
6. Poletti E, Zappelli F, Ruggeri A, Grisan E (2012) A review of thersholding strategies applied to human chromosome segmentation. Comput Methods Programs Biomed 108:679–688
7. Bhandari D, Pal NR, Majumder DD (1992) Fuzzy divergence, probability measure of fuzzy event and image thresholding. Inf Sci 13:857–867
8. Fan S, Yang S, He P, Nie H (2011) Infrared electric image thresholding using two dimensional fuzzy entropy. Energy Procedia 12:411–419
9. Bhatia PK, Singh S (2013) A new measure of fuzzy directed divergence and its application in image segmentation. Int J Intell Syst Appl 4:81–89
10. Ghosh M, Das D, Chakraborty C, Ray AK (2010) Automated leukocyte recognition using fuzzy divergence. Micron 41:840–846
11. Shore JE, Gray RM (1982) Minimization cross–entropy pattern classification and cluster analysis. IEEE Trans Pattern Anal Mach Intell 4(1):11–17

Chapter 6
Generalized Hellinger's Fuzzy Divergence Measure and Its Applications

In real-life situations, the human thinking style involves subjectivity which introduces vagueness in decision-making. This vagueness or uncertainty is easily handled by fuzzy set theory. This chapter deals with the issue of Hellinger's generalized measure of discrimination in fuzzy environment. Multi-criteria decision-making (MCDM) has been one of the rapidly developing areas based on the altering in the commercial area. For this, decision-maker requires a decision support to decide among different options or alternatives by surpassing the less suitable options quickly. At the managerial and industrial point Duckstein and Opricovic [3] considered multiple-criteria decision-making as a compound and dynamic process. Hwang and Yoon [7] and Dyer et al. [4] precisely presented the method of weights and ranking of the criteria in the classical MCDM methods. Zeleny [11] and Hwang and Yoon [7] presented a survey of the MCDM methods based on the concept that the preferred alternative should have a smaller distance from the positive-ideal solution and the largest distance from the negative-ideal solution.

Bellman and Zadeh [1] introduced the theory of solving MCDM problems in a fuzzy environment which is acknowledged as fuzzy multi-criteria decision-making (FMCDM) problems. Erginel et al. [5] analyzed that the fuzzy theory is often used in the study of multi-criteria decision-making approaches since MCDM problems are mainly depend on individual views and qualitative information.

Hellinger's measure of discrimination was first introduced by Hellinger [6]. Further, Taneja [10] proposed a generalized Hellinger's discrimination given by

$$h_t(P,Q)=\sum_{i=1}^{n}\frac{\left(\sqrt{p_i}-\sqrt{q_i}\right)^{2(t+1)}}{\left(\sqrt{p_i q_i}\right)^{t}},\quad (P,Q)\in\Gamma_n\times\Gamma_n,\ t\in N.$$

$$\text{where }\Gamma_n=\left\{P=(p_1,p_2,\ldots,p_n)\Big/ p_i>0,\sum_{i=1}^{n}p_i=1\right\},\quad n\geq 2. \tag{6.1}$$

A. Ohlan and R. Ohlan, *Generalizations of Fuzzy Information Measures*,
DOI 10.1007/978-3-319-45928-8_6

In this chapter, we propose a new parametric generalized Hellinger's divergence measure in fuzzy environment corresponding to the generalized Hellinger's measure of discrimination (6.1) of Taneja [10] which provides a flexible approach to further leverage of choice to the user. It may be observed that the potential, strength and efficiency of this new generalized Hellinger's fuzzy divergence measure exist in its properties and the practical application to multi-criteria decision-making problems and medical diagnosis.

In Sect. 6.1 we introduce a generalized Hellinger's fuzzy divergence measure with the proof of its validity. Some interesting properties of the proposed measure between different fuzzy sets are studied in Sect. 6.2. In this process, Sect. 6.3 provides a method for solving the problem related to multi-criteria decision-making. In the same section, the applications of the proposed generalization of Hellinger's fuzzy divergence measure are illustrated in multi-criteria decision-making problem and medical diagnosis. Finally, concluding remarks are drawn in Sect. 6.4.

6.1 New Generalized Fuzzy Hellinger's Divergence Measure

We now propose a generalized measure of Hellinger's divergence between two fuzzy sets A, B of universe of discourse $X = \{x_1, x_2, \ldots, x_n\}$ having the membership values $\mu_A(x_i)$, $\mu_B(x_i) \in (0, 1)$ corresponding to Taneja [10] generalized Hellinger's divergence measure (6.1) given by

$$h_t(A, B) = \sum_{i=1}^{n} \left[\frac{\left(\sqrt{\mu_A(x_i)} - \sqrt{\mu_B(x_i)}\right)^{2(t+1)}}{\left(\sqrt{\mu_A(x_i)\mu_B(x_i)}\right)^{t}} + \frac{\left(\sqrt{1-\mu_A(x_i)} - \sqrt{1-\mu_B(x_i)}\right)^{2(t+1)}}{\left(\sqrt{(1-\mu_A(x_i))(1-\mu_B(x_i))}\right)^{t}} \right], \quad t \in N. \tag{6.2}$$

Theorem 6.1 *$h_t(A, B)$ is the valid measure of fuzzy divergence.*

Proof It is clear from (6.2) that

(i) $h_t(A, B) \geq 0$
(ii) $h_t(A, B) = 0$ if $\mu_A(x_i) = \mu_B(x_i)$, $\forall i = 1, 2, \ldots, n$.
(iii) We now check the convexity of $h_t(A, B)$.

$$\frac{\partial h_t(A,B)}{\partial \mu_A(x_i)}$$
$$= \frac{(2t+2)\left(\sqrt{\mu_A(x_i)} - \sqrt{\mu_B(x_i)}\right)^{(2t+1)}}{(\mu_A(x_i)\mu_B(x_i))^{t/2}} \frac{1}{2\sqrt{\mu_A(x_i)}}$$
$$+ \left(-\frac{t}{2}\right) \frac{\left(\sqrt{\mu_A(x_i)} - \sqrt{\mu_B(x_i)}\right)^{(2t+2)}}{(\mu_A(x_i)\mu_B(x_i))^{-\frac{t}{2}-1}} \mu_B(x_i)$$
$$+ 2(t+1) \frac{\left(\sqrt{1-\mu_A(x_i)} - \sqrt{1-\mu_B(x_i)}\right)^{(2t+1)}}{\left(\sqrt{(1-\mu_A(x_i))(1-\mu_B(x_i))}\right)^t} \frac{1}{2\sqrt{1-\mu_A(x_i)}}(-1)$$
$$- \frac{t}{2} \frac{\left(\sqrt{1-\mu_A(x_i)} - \sqrt{1-\mu_B(x_i)}\right)^{(2t+2)}}{((1-\mu_A(x_i))(1-\mu_B(x_i)))^{\frac{t}{2}+1}} [-(1-\mu_B(x_i))]$$
$$= (t+1)\left[\frac{\left(\sqrt{\mu_A(x_i)} - \sqrt{\mu_B(x_i)}\right)^{(2t+1)}}{\sqrt{\mu_A(x_i)}\left(\sqrt{\mu_A(x_i)\mu_B(x_i)}\right)^t} - \frac{\left(\sqrt{1-\mu_A(x_i)} - \sqrt{1-\mu_B(x_i)}\right)^{(2t+1)}}{\sqrt{1-\mu_A(x_i)}\left(\sqrt{(1-\mu_A(x_i))(1-\mu_B(x_i))}\right)^t}\right]$$
$$- \frac{t}{2}\left[\frac{\left(\sqrt{\mu_A(x_i)} - \sqrt{\mu_B(x_i)}\right)^{(2t+2)}}{\mu_A(x_i)\left(\sqrt{\mu_A(x_i)\mu_B(x_i)}\right)^t} - \frac{\left(\sqrt{1-\mu_A(x_i)} - \sqrt{1-\mu_B(x_i)}\right)^{(2t+2)}}{(1-\mu_A(x_i))\left(\sqrt{(1-\mu_A(x_i))(1-\mu_B(x_i))}\right)^t}\right]$$

$$\frac{\partial^2 h_t(A,B)}{\partial \mu_A^2(x_i)}$$
$$= \frac{(t+1)(2t+1)}{2}\left[\frac{\left(\sqrt{\mu_A(x_i)} - \sqrt{\mu_B(x_i)}\right)^{2t}}{\mu_A(x_i)\left(\sqrt{\mu_A(x_i)\mu_B(x_i)}\right)^t} + \frac{\left(\sqrt{1-\mu_A(x_i)} - \sqrt{1-\mu_B(x_i)}\right)^{2t}}{(1-\mu_A(x_i))\left(\sqrt{(1-\mu_A(x_i))(1-\mu_B(x_i))}\right)^t}\right]$$
$$- \frac{(t+1)(2t+1)}{2}\left[\frac{\left(\sqrt{\mu_A(x_i)} - \sqrt{\mu_B(x_i)}\right)^{2t+1}}{\mu_A^{3/2}(x_i)\left(\sqrt{\mu_A(x_i)\mu_B(x_i)}\right)^t} + \frac{\left(\sqrt{1-\mu_A(x_i)} - \sqrt{1-\mu_B(x_i)}\right)^{2t+1}}{(1-\mu_A(x_i))^{3/2}\left(\sqrt{(1-\mu_A(x_i))(1-\mu_B(x_i))}\right)^t}\right]$$
$$+ \frac{t(t+2)}{4}\left[\frac{\left(\sqrt{\mu_A(x_i)} - \sqrt{\mu_B(x_i)}\right)^{2t+2}}{\mu_A^2(x_i)\left(\sqrt{\mu_A(x_i)\mu_B(x_i)}\right)^t} + \frac{\left(\sqrt{1-\mu_A(x_i)} - \sqrt{1-\mu_B(x_i)}\right)^{2t+2}}{(1-\mu_A(x_i))^2\left(\sqrt{(1-\mu_A(x_i))(1-\mu_B(x_i))}\right)^t}\right] > 0, \quad t \in N.$$

Similarly, $\frac{\partial^2 h_t(A,B)}{\partial \mu_B^2(x_i)} > 0, t \in N$.

Thus $h_t(A,B)$ is a convex function of fuzzy sets A and B and hence in view of the definition of fuzzy divergence measure of Bhandari and Pal [2] $h_t(A,B)$ is a valid measure of fuzzy divergence. Moreover, we can easily check for measure $h_t(A,B)$

in (6.2) that $h_t(A,B) = h_t(B,A)$. Hence the defined measure $h_t(A,B)$ is a valid measure of fuzzy symmetric divergence.

In particular,

For $t = 0$, $h_t(A,B)$ reduces to $2h(A,B)$ where $h(A,B)$ is fuzzy Hellinger's divergence measure of Singh and Tomar [9].

6.2 Properties of Generalized Fuzzy Hellinger's Divergence Measure

In this section, we provide some more properties of the proposed generalized measure of Hellinger's fuzzy divergence (6.2) in the following theorems. While proving these theorems we consider the separation of X into two parts X_1 and X_2 as

$$X_1 = \{x/x \in X, \mu_A(x_i) \geq \mu_B(x_i)\} \tag{6.3}$$

$$\text{and}\, X_2 = \{x/x \in X, \mu_A(x_i) < \mu_B(x_i)\}. \tag{6.4}$$

Theorem 6.2

(a) $h_t(A \cup B, A) + h_t(A \cap B, A) = h_t(A,B)$.
(b) $h_t(\overline{A \cup B}, \overline{A \cap B}) = h_t(\overline{A} \cap \overline{B}, \overline{A} \cup \overline{B}) = h_t(A,B)$.
(c) $h_t(A, \overline{A}) = h_t(\overline{A}, A)$.

Proof 6.2(a)

$$\begin{aligned}
&h_t(A \cup B, A) + h_t(A \cap B, A) \\
&= \sum_{i=1}^{n} \left[\frac{\left(\sqrt{\mu_{A\cup B}(x_i)} - \sqrt{\mu_A(x_i)}\right)^{2(t+1)}}{\left(\sqrt{\mu_{A\cup B}(x_i)\mu_A(x_i)}\right)^t} + \frac{\left(\sqrt{1-\mu_{A\cup B}(x_i)} - \sqrt{1-\mu_A(x_i)}\right)^{2(t+1)}}{\left(\sqrt{(1-\mu_{A\cup B}(x_i))(1-\mu_A(x_i))}\right)^t} \right] \\
&\quad + \sum_{i=1}^{n} \left[\frac{\left(\sqrt{\mu_{A\cap B}(x_i)} - \sqrt{\mu_A(x_i)}\right)^{2(t+1)}}{\left(\sqrt{\mu_{A\cap B}(x_i)\mu_A(x_i)}\right)^t} + \frac{\left(\sqrt{1-\mu_{A\cap B}(x_i)} - \sqrt{1-\mu_A(x_i)}\right)^{2(t+1)}}{\left(\sqrt{(1-\mu_{A\cap B}(x_i))(1-\mu_A(x_i))}\right)^t} \right] \\
&= \sum_{X_2} \left[\frac{\left(\sqrt{\mu_B(x_i)} - \sqrt{\mu_A(x_i)}\right)^{2(t+1)}}{\left(\sqrt{\mu_A(x_i)\mu_B(x_i)}\right)^t} + \frac{\left(\sqrt{1-\mu_B(x_i)} - \sqrt{1-\mu_A(x_i)}\right)^{2(t+1)}}{\left(\sqrt{(1-\mu_A(x_i))(1-\mu_B(x_i))}\right)^t} \right] \\
&\quad + \sum_{X_1} \left[\frac{\left(\sqrt{\mu_B(x_i)} - \sqrt{\mu_A(x_i)}\right)^{2(t+1)}}{\left(\sqrt{\mu_A(x_i)\mu_B(x_i)}\right)^t} + \frac{\left(\sqrt{1-\mu_B(x_i)} - \sqrt{1-\mu_A(x_i)}\right)^{2(t+1)}}{\left(\sqrt{(1-\mu_A(x_i))(1-\mu_B(x_i))}\right)^t} \right] \\
&= \sum_{i=1}^{n} \left[\frac{\left(\sqrt{\mu_A(x_i)} - \sqrt{\mu_B(x_i)}\right)^{2(t+1)}}{\left(\sqrt{\mu_A(x_i)\mu_B(x_i)}\right)^t} + \frac{\left(\sqrt{1-\mu_A(x_i)} - \sqrt{1-\mu_B(x_i)}\right)^{2(t+1)}}{\left(\sqrt{(1-\mu_A(x_i))(1-\mu_B(x_i))}\right)^t} \right] \\
&= h_t(A,B).
\end{aligned}$$

Hence, 6.2(a) holds.
6.2(b)

$$h_t(\overline{A \cup B}, \overline{A \cap B})$$
$$= \sum_{i=1}^{n}\left[\frac{\left(\sqrt{\mu_{\overline{A\cup B}}(x_i)} - \sqrt{\mu_{\overline{A\cap B}}(x_i)}\right)^{2(t+1)}}{\left(\sqrt{\mu_{\overline{A\cup B}}(x_i)\mu_{\overline{A\cap B}}(x_i)}\right)^t} + \frac{\left(\sqrt{1-\mu_{\overline{A\cup B}}(x_i)} - \sqrt{1-\mu_{\overline{A\cap B}}(x_i)}\right)^{2(t+1)}}{\left(\sqrt{(1-\mu_{\overline{A\cup B}}(x_i))(1-\mu_{\overline{A\cap B}}(x_i))}\right)^t}\right]$$
$$= \sum_{i=1}^{n}\left[\frac{\left(\sqrt{1-\mu_{A\cup B}(x_i)} - \sqrt{1-\mu_{A\cap B}(x_i)}\right)^{2(t+1)}}{\left(\sqrt{(1-\mu_{A\cup B}(x_i))(1-\mu_{A\cap B}(x_i))}\right)^t} + \frac{\left(\sqrt{\mu_{A\cup B}(x_i)} - \sqrt{\mu_{A\cap B}(x_i)}\right)^{2(t+1)}}{\left(\sqrt{\mu_{A\cup B}(x_i)\mu_{A\cap B}(x_i)}\right)^t}\right]$$
$$= \sum_{X_1}\left[\frac{\left(\sqrt{1-\mu_A(x_i)} - \sqrt{1-\mu_B(x_i)}\right)^{2(t+1)}}{\left(\sqrt{(1-\mu_A(x_i))(1-\mu_B(x_i))}\right)^t} + \frac{\left(\sqrt{\mu_A(x_i)} - \sqrt{\mu_B(x_i)}\right)^{2(t+1)}}{\left(\sqrt{\mu_A(x_i)\mu_B(x_i)}\right)^t}\right]$$
$$+ \sum_{X_2}\left[\frac{\left(\sqrt{1-\mu_B(x_i)} - \sqrt{1-\mu_A(x_i)}\right)^{2(t+1)}}{\left(\sqrt{(1-\mu_B(x_i))(1-\mu_A(x_i))}\right)^t} + \frac{\left(\sqrt{\mu_B(x_i)} - \sqrt{\mu_A(x_i)}\right)^{2(t+1)}}{\left(\sqrt{\mu_B(x_i)\mu_A(x_i)}\right)^t}\right]$$
$$= h_t(A, B).$$

Now

$$h_t(\overline{A} \cap \overline{B}, \overline{A} \cup \overline{B})$$
$$= \sum_{i=1}^{n}\left[\frac{\left(\sqrt{\mu_{\overline{A}\cap\overline{B}}(x_i)} - \sqrt{\mu_{\overline{A}\cup\overline{B}}(x_i)}\right)^{2(t+1)}}{\left(\sqrt{\mu_{\overline{A}\cap\overline{B}}(x_i)\mu_{\overline{A}\cup\overline{B}}(x_i)}\right)^t} + \frac{\left(\sqrt{1-\mu_{\overline{A}\cap\overline{B}}(x_i)} - \sqrt{1-\mu_{\overline{A}\cup\overline{B}}(x_i)}\right)^{2(t+1)}}{\left(\sqrt{(1-\mu_{\overline{A}\cap\overline{B}}(x_i))(1-\mu_{\overline{A}\cup\overline{B}}(x_i))}\right)^t}\right]$$
$$= \sum_{X_1}\left[\frac{\left(\sqrt{\mu_{\overline{A}}(x_i)} - \sqrt{\mu_{\overline{B}}(x_i)}\right)^{2(t+1)}}{\left(\sqrt{\mu_{\overline{A}}(x_i)\mu_{\overline{B}}(x_i)}\right)^t} + \frac{\left(\sqrt{1-\mu_{\overline{A}}(x_i)} - \sqrt{1-\mu_{\overline{B}}(x_i)}\right)^{2(t+1)}}{\left(\sqrt{(1-\mu_{\overline{A}}(x_i))(1-\mu_{\overline{B}}(x_i))}\right)^t}\right]$$
$$+ \sum_{X_2}\left[\frac{\left(\sqrt{\mu_{\overline{B}}(x_i)} - \sqrt{\mu_{\overline{A}}(x_i)}\right)^{2(t+1)}}{\left(\sqrt{\mu_{\overline{B}}(x_i)\mu_{\overline{A}}(x_i)}\right)^t} + \frac{\left(\sqrt{1-\mu_{\overline{B}}(x_i)} - \sqrt{1-\mu_{\overline{A}}(x_i)}\right)^{2(t+1)}}{\left(\sqrt{(1-\mu_{\overline{B}}(x_i))(1-\mu_{\overline{A}}(x_i))}\right)^t}\right]$$
$$= \sum_{X_1}\left[\frac{\left(\sqrt{1-\mu_A(x_i)} - \sqrt{1-\mu_B(x_i)}\right)^{2(t+1)}}{\left(\sqrt{(1-\mu_A(x_i))(1-\mu_B(x_i))}\right)^t} + \frac{\left(\sqrt{\mu_A(x_i)} - \sqrt{\mu_B(x_i)}\right)^{2(t+1)}}{\left(\sqrt{\mu_A(x_i)\mu_B(x_i)}\right)^t}\right]$$
$$+ \sum_{X_2}\left[\frac{\left(\sqrt{1-\mu_B(x_i)} - \sqrt{1-\mu_A(x_i)}\right)^{2(t+1)}}{\left(\sqrt{(1-\mu_B(x_i))(1-\mu_A(x_i))}\right)^t} + \frac{\left(\sqrt{\mu_B(x_i)} - \sqrt{\mu_A(x_i)}\right)^{2(t+1)}}{\left(\sqrt{\mu_B(x_i)\mu_A(x_i)}\right)^t}\right]$$
$$= h_t(A, B).$$

Hence, $h_t(\overline{A \cup B}, \overline{A \cap B}) = h_t(\overline{A} \cap \overline{B}, \overline{A} \cup \overline{B}) = h_t(A, B)$.

Hence, 6.2(b) holds.
6.2(c)

$$h_t(A,\overline{A}) = \sum_{i=1}^{n}\left[\frac{\left(\sqrt{\mu_A(x_i)} - \sqrt{\mu_{\overline{A}}(x_i)}\right)^{2(t+1)}}{\left(\sqrt{\mu_A(x_i)\mu_{\overline{A}}(x_i)}\right)^t} + \frac{\left(\sqrt{1-\mu_A(x_i)} - \sqrt{1-\mu_{\overline{A}}(x_i)}\right)^{2(t+1)}}{\left(\sqrt{(1-\mu_A(x_i))(1-\mu_{\overline{A}}(x_i))}\right)^t}\right]$$
$$= \sum_{i=1}^{n}\left[\frac{\left(\sqrt{\mu_A(x_i)} - \sqrt{1-\mu_A(x_i)}\right)^{2(t+1)}}{\left(\sqrt{\mu_A(x_i)(1-\mu_A(x_i))}\right)^t} + \frac{\left(\sqrt{1-\mu_A(x_i)} - \sqrt{\mu_A(x_i)}\right)^{2(t+1)}}{\left(\sqrt{(1-\mu_A(x_i))\mu_A(x_i)}\right)^t}\right]$$
$$= \sum_{i=1}^{n}\left[\frac{2\left(\sqrt{\mu_A(x_i)} - \sqrt{1-\mu_A(x_i)}\right)^{2(t+1)}}{\left(\sqrt{\mu_A(x_i)(1-\mu_A(x_i))}\right)^t}\right]$$
$$h_t(\overline{A},A) = \sum_{i=1}^{n}\left[\frac{\left(\sqrt{\mu_{\overline{A}}(x_i)} - \sqrt{\mu_A(x_i)}\right)^{2(t+1)}}{\left(\sqrt{\mu_{\overline{A}}(x_i)\mu_A(x_i)}\right)^t} + \frac{\left(\sqrt{1-\mu_{\overline{A}}(x_i)} - \sqrt{1-\mu_A(x_i)}\right)^{2(t+1)}}{\left(\sqrt{(1-\mu_{\overline{A}}(x_i))(1-\mu_A(x_i))}\right)^t}\right]$$
$$= \sum_{i=1}^{n}\left[\frac{\left(\sqrt{1-\mu_A(x_i)} - \sqrt{\mu_A(x_i)}\right)^{2(t+1)}}{\left(\sqrt{(1-\mu_A(x_i))\mu_A(x_i)}\right)^t} + \frac{\left(\sqrt{\mu_A(x_i)} - \sqrt{(1-\mu_A(x_i))}\right)^{2(t+1)}}{\left(\sqrt{\mu_A(x_i)(1-\mu_A(x_i))}\right)^t}\right]$$
$$= \sum_{i=1}^{n}\left[\frac{2\left(\sqrt{\mu_A(x_i)} - \sqrt{1-\mu_A(x_i)}\right)^{2(t+1)}}{\left(\sqrt{\mu_A(x_i)(1-\mu_A(x_i))}\right)^t}\right]$$
$$= h_t(A,\overline{A}).$$

Hence, $h_t(A,\overline{A}) = h_t(\overline{A},A)$.
Hence, 6.2(c) holds.

Theorem 6.3

(a) $h_t(A\cup B,C) + h_t(A\cap B,C) = h_t(A,C) + h_t(B,C)$.
(b) $h_t(A,A\cup B) = h_t(B,A\cap B)$.
(c) $h_t(A,A\cap B) = h_t(B,A\cup B)$.

Proof 6.3(a)

$$h_t(A\cup B,C) + h_t(A\cap B,C)$$
$$= \sum_{i=1}^{n}\left[\frac{\left(\sqrt{\mu_{A\cup B}(x_i)} - \sqrt{\mu_C(x_i)}\right)^{2(t+1)}}{\left(\sqrt{\mu_{A\cup B}(x_i)\mu_C(x_i)}\right)^t} + \frac{\left(\sqrt{1-\mu_{A\cup B}(x_i)} - \sqrt{1-\mu_C(x_i)}\right)^{2(t+1)}}{\left(\sqrt{(1-\mu_{A\cup B}(x_i))(1-\mu_C(x_i))}\right)^t}\right]$$
$$+ \sum_{i=1}^{n}\left[\frac{\left(\sqrt{\mu_{A\cap B}(x_i)} - \sqrt{\mu_C(x_i)}\right)^{2(t+1)}}{\left(\sqrt{\mu_{A\cap B}(x_i)\mu_C(x_i)}\right)^t} + \frac{\left(\sqrt{1-\mu_{A\cap B}(x_i)} - \sqrt{1-\mu_C(x_i)}\right)^{2(t+1)}}{\left(\sqrt{(1-\mu_{A\cap B}(x_i))(1-\mu_C(x_i))}\right)^t}\right]$$
$$= \sum_{X_1}\left[\frac{\left(\sqrt{\mu_A(x_i)} - \sqrt{\mu_C(x_i)}\right)^{2(t+1)}}{\left(\sqrt{\mu_A(x_i)\mu_C(x_i)}\right)^t} + \frac{\left(\sqrt{1-\mu_A(x_i)} - \sqrt{1-\mu_C(x_i)}\right)^{2(t+1)}}{\left(\sqrt{(1-\mu_A(x_i))(1-\mu_C(x_i))}\right)^t}\right]$$
$$+ \sum_{X_2}\left[\frac{\left(\sqrt{\mu_B(x_i)} - \sqrt{\mu_C(x_i)}\right)^{2(t+1)}}{\left(\sqrt{\mu_B(x_i)\mu_C(x_i)}\right)^t} + \frac{\left(\sqrt{1-\mu_B(x_i)} - \sqrt{1-\mu_C(x_i)}\right)^{2(t+1)}}{\left(\sqrt{(1-\mu_B(x_i))(1-\mu_C(x_i))}\right)^t}\right]$$

$$+\sum_{X_1}\left[\frac{\left(\sqrt{\mu_B(x_i)}-\sqrt{\mu_C(x_i)}\right)^{2(t+1)}}{\left(\sqrt{\mu_B(x_i)\mu_C(x_i)}\right)^t}+\frac{\left(\sqrt{1-\mu_B(x_i)}-\sqrt{1-\mu_C(x_i)}\right)^{2(t+1)}}{\left(\sqrt{(1-\mu_B(x_i))(1-\mu_C(x_i))}\right)^t}\right]$$
$$+\sum_{X_2}\left[\frac{\left(\sqrt{\mu_A(x_i)}-\sqrt{\mu_C(x_i)}\right)^{2(t+1)}}{\left(\sqrt{\mu_A(x_i)\mu_C(x_i)}\right)^t}+\frac{\left(\sqrt{1-\mu_A(x_i)}-\sqrt{1-\mu_C(x_i)}\right)^{2(t+1)}}{\left(\sqrt{(1-\mu_A(x_i))(1-\mu_C(x_i))}\right)^t}\right]$$
$$=\sum_{i=1}^{n}\left[\frac{\left(\sqrt{\mu_A(x_i)}-\sqrt{\mu_C(x_i)}\right)^{2(t+1)}}{\left(\sqrt{\mu_A(x_i)\mu_C(x_i)}\right)^t}+\frac{\left(\sqrt{1-\mu_A(x_i)}-\sqrt{1-\mu_C(x_i)}\right)^{2(t+1)}}{\left(\sqrt{(1-\mu_A(x_i))(1-\mu_C(x_i))}\right)^t}\right]$$
$$+\sum_{i=1}^{n}\left[\frac{\left(\sqrt{\mu_B(x_i)}-\sqrt{\mu_C(x_i)}\right)^{2(t+1)}}{\left(\sqrt{\mu_B(x_i)\mu_C(x_i)}\right)^t}+\frac{\left(\sqrt{1-\mu_B(x_i)}-\sqrt{1-\mu_C(x_i)}\right)^{2(t+1)}}{\left(\sqrt{(1-\mu_B(x_i))(1-\mu_C(x_i))}\right)^t}\right]$$
$$=h_t(A,C)+h_t(B,C).$$

Hence, 6.3(a) holds.
6.3(b)

$$h_t(A,A\cup B)$$
$$=\sum_{i=1}^{n}\left[\frac{\left(\sqrt{\mu_A(x_i)}-\sqrt{\mu_{A\cup B}(x_i)}\right)^{2(t+1)}}{\left(\sqrt{\mu_A(x_i)\mu_{A\cup B}(x_i)}\right)^t}+\frac{\left(\sqrt{1-\mu_A(x_i)}-\sqrt{1-\mu_{A\cup B}(x_i)}\right)^{2(t+1)}}{\left(\sqrt{(1-\mu_A(x_i))(1-\mu_{A\cup B}(x_i))}\right)^t}\right]$$
$$=\sum_{X_2}\left[\frac{\left(\sqrt{\mu_A(x_i)}-\sqrt{\mu_B(x_i)}\right)^{2(t+1)}}{\left(\sqrt{\mu_A(x_i)\mu_B(x_i)}\right)^t}+\frac{\left(\sqrt{1-\mu_A(x_i)}-\sqrt{1-\mu_B(x_i)}\right)^{2(t+1)}}{\left(\sqrt{(1-\mu_A(x_i))(1-\mu_B(x_i))}\right)^t}\right]$$

Now

$$h_t(B,A\cap B)$$
$$=\sum_{i=1}^{n}\left[\frac{\left(\sqrt{\mu_B(x_i)}-\sqrt{\mu_{A\cap B}(x_i)}\right)^{2(t+1)}}{\left(\sqrt{\mu_B(x_i)\mu_{A\cap B}(x_i)}\right)^t}+\frac{\left(\sqrt{1-\mu_B(x_i)}-\sqrt{1-\mu_{A\cap B}(x_i)}\right)^{2(t+1)}}{\left(\sqrt{(1-\mu_B(x_i))(1-\mu_{A\cap B}(x_i))}\right)^t}\right]$$
$$=\sum_{X_2}\left[\frac{\left(\sqrt{\mu_B(x_i)}-\sqrt{\mu_A(x_i)}\right)^{2(t+1)}}{\left(\sqrt{\mu_B(x_i)\mu_A(x_i)}\right)^t}+\frac{\left(\sqrt{1-\mu_B(x_i)}-\sqrt{1-\mu_A(x_i)}\right)^{2(t+1)}}{\left(\sqrt{(1-\mu_B(x_i))(1-\mu_A(x_i))}\right)^t}\right]$$
$$=h_t(A,A\cup B).$$

Hence, 6.3(b) holds.
6.3(c)

$$h_t(A,A\cap B)$$
$$=\sum_{i=1}^{n}\left[\frac{\left(\sqrt{\mu_A(x_i)}-\sqrt{\mu_{A\cap B}(x_i)}\right)^{2(t+1)}}{\left(\sqrt{\mu_A(x_i)\mu_{A\cap B}(x_i)}\right)^t}+\frac{\left(\sqrt{1-\mu_A(x_i)}-\sqrt{1-\mu_{A\cap B}(x_i)}\right)^{2(t+1)}}{\left(\sqrt{(1-\mu_A(x_i))(1-\mu_{A\cap B}(x_i))}\right)^t}\right]$$
$$=\sum_{X_1}\left[\frac{\left(\sqrt{\mu_A(x_i)}-\sqrt{\mu_B(x_i)}\right)^{2(t+1)}}{\left(\sqrt{\mu_A(x_i)\mu_B(x_i)}\right)^t}+\frac{\left(\sqrt{1-\mu_A(x_i)}-\sqrt{1-\mu_B(x_i)}\right)^{2(t+1)}}{\left(\sqrt{(1-\mu_A(x_i))(1-\mu_B(x_i))}\right)^t}\right]$$

Now

$$
\begin{aligned}
&h_t(B, A\cup B)\\
&=\sum_{i=1}^{n}\left[\frac{\left(\sqrt{\mu_B(x_i)}-\sqrt{\mu_{A\cup B}(x_i)}\right)^{2(t+1)}}{\left(\sqrt{\mu_B(x_i)\mu_{A\cup B}(x_i)}\right)^{t}}+\frac{\left(\sqrt{1-\mu_B(x_i)}-\sqrt{1-\mu_{A\cup B}(x_i)}\right)^{2(t+1)}}{\left(\sqrt{(1-\mu_B(x_i))(1-\mu_{A\cup B}(x_i))}\right)^{t}}\right]\\
&=\sum_{X_1}\left[\frac{\left(\sqrt{\mu_B(x_i)}-\sqrt{\mu_A(x_i)}\right)^{2(t+1)}}{\left(\sqrt{\mu_B(x_i)\mu_A(x_i)}\right)^{t}}+\frac{\left(\sqrt{1-\mu_B(x_i)}-\sqrt{1-\mu_A(x_i)}\right)^{2(t+1)}}{\left(\sqrt{(1-\mu_B(x_i))(1-\mu_A(x_i))}\right)^{t}}\right]\\
&=h_t(A, A\cap B).
\end{aligned}
$$

Hence, 6.3(c) holds.

Theorem 6.4

(a) $h_t(A\cup B, A\cap B)=h_t(A,B)$.
(b) $h_t(\overline{A},\overline{B})=h_t(A,B)$.
(c) $h_t(A,\overline{B})=h_t(\overline{A},B)$.
(d) $h_t(A,B)+h_t(\overline{A},B)=h_t(\overline{A},\overline{B})+h_t(A,\overline{B})$.

Proof 6.4(a)

$$
\begin{aligned}
&h_t(A\cup B, A\cap B)\\
&=\sum_{i=1}^{n}\left[\frac{\left(\sqrt{\mu_{A\cup B}(x_i)}-\sqrt{\mu_{A\cap B}(x_i)}\right)^{2(t+1)}}{\left(\sqrt{\mu_{A\cup B}(x_i)\mu_{A\cap B}(x_i)}\right)^{t}}+\frac{\left(\sqrt{1-\mu_{A\cup B}(x_i)}-\sqrt{1-\mu_{A\cap B}(x_i)}\right)^{2(t+1)}}{\left(\sqrt{(1-\mu_{A\cup B}(x_i))(1-\mu_{A\cap B}(x_i))}\right)^{t}}\right]\\
&=\sum_{X_1}\left[\frac{\left(\sqrt{\mu_A(x_i)}-\sqrt{\mu_B(x_i)}\right)^{2(t+1)}}{\left(\sqrt{\mu_A(x_i)\mu_B(x_i)}\right)^{t}}+\frac{\left(\sqrt{1-\mu_A(x_i)}-\sqrt{1-\mu_B(x_i)}\right)^{2(t+1)}}{\left(\sqrt{(1-\mu_A(x_i))(1-\mu_B(x_i))}\right)^{t}}\right]\\
&\quad+\sum_{X_2}\left[\frac{\left(\sqrt{\mu_B(x_i)}-\sqrt{\mu_A(x_i)}\right)^{2(t+1)}}{\left(\sqrt{\mu_B(x_i)\mu_A(x_i)}\right)^{t}}+\frac{\left(\sqrt{1-\mu_B(x_i)}-\sqrt{1-\mu_A(x_i)}\right)^{2(t+1)}}{\left(\sqrt{(1-\mu_B(x_i))(1-\mu_A(x_i))}\right)^{t}}\right]s\\
&=\sum_{i=1}^{n}\left[\frac{\left(\sqrt{\mu_A(x_i)}-\sqrt{\mu_B(x_i)}\right)^{2(t+1)}}{\left(\sqrt{\mu_A(x_i)\mu_B(x_i)}\right)^{t}}+\frac{\left(\sqrt{1-\mu_A(x_i)}-\sqrt{1-\mu_B(x_i)}\right)^{2(t+1)}}{\left(\sqrt{(1-\mu_A(x_i))(1-\mu_B(x_i))}\right)^{t}}\right]\\
&=h_t(A,B).
\end{aligned}
$$

Hence, 6.4(a) holds.

6.4(b)

$$
\begin{aligned}
h_t(\overline{A},\overline{B}) &= \sum_{i=1}^{n}\left[\frac{\left(\sqrt{\mu_{\overline{A}}(x_i)}-\sqrt{\mu_{\overline{B}}(x_i)}\right)^{2(t+1)}}{\left(\sqrt{\mu_{\overline{A}}(x_i)\mu_{\overline{B}}(x_i)}\right)^{t}}+\frac{\left(\sqrt{1-\mu_{\overline{A}}(x_i)}-\sqrt{1-\mu_{\overline{B}}(x_i)}\right)^{2(t+1)}}{\left(\sqrt{(1-\mu_{\overline{A}}(x_i))(1-\mu_{\overline{B}}(x_i))}\right)^{t}}\right]\\
&= \sum_{i=1}^{n}\left[\frac{\left(\sqrt{(1-\mu_A(x_i))}-\sqrt{(1-\mu_B(x_i))}\right)^{2(t+1)}}{\left(\sqrt{(1-\mu_A(x_i))(1-\mu_B(x_i))}\right)^{t}}+\frac{\left(\sqrt{\mu_A(x_i)}-\sqrt{\mu_B(x_i)}\right)^{2(t+1)}}{\left(\sqrt{\mu_A(x_i)\mu_B(x_i)}\right)^{t}}\right]\\
&= h_t(A,B).
\end{aligned}
$$

Hence, 6.4(b) holds.

6.4(c)

$$
\begin{aligned}
h_t(A,\overline{B}) &= \sum_{i=1}^{n}\left[\frac{\left(\sqrt{\mu_A(x_i)}-\sqrt{\mu_{\overline{B}}(x_i)}\right)^{2(t+1)}}{\left(\sqrt{\mu_A(x_i)\mu_{\overline{B}}(x_i)}\right)^{t}}+\frac{\left(\sqrt{1-\mu_A(x_i)}-\sqrt{1-\mu_{\overline{B}}(x_i)}\right)^{2(t+1)}}{\left(\sqrt{(1-\mu_A(x_i))(1-\mu_{\overline{B}}(x_i))}\right)^{t}}\right]\\
&= \sum_{i=1}^{n}\left[\frac{\left(\sqrt{\mu_A(x_i)}-\sqrt{(1-\mu_B(x_i))}\right)^{2(t+1)}}{\left(\sqrt{\mu_A(x_i)(1-\mu_B(x_i))}\right)^{t}}+\frac{\left(\sqrt{(1-\mu_A(x_i))}-\sqrt{\mu_B(x_i)}\right)^{2(t+1)}}{\left(\sqrt{(1-\mu_A(x_i))\mu_B(x_i)}\right)^{t}}\right]\\
h_t(\overline{A},B) &= \sum_{i=1}^{n}\left[\frac{\left(\sqrt{\mu_{\overline{A}}(x_i)}-\sqrt{\mu_B(x_i)}\right)^{2(t+1)}}{\left(\sqrt{\mu_{\overline{A}}(x_i)\mu_B(x_i)}\right)^{t}}+\frac{\left(\sqrt{(1-\mu_{\overline{A}}(x_i))}-\sqrt{(1-\mu_B(x_i))}\right)^{2(t+1)}}{\left(\sqrt{(1-\mu_{\overline{A}}(x_i))(1-\mu_B(x_i))}\right)^{t}}\right]\\
&= \sum_{i=1}^{n}\left[\frac{\left(\sqrt{(1-\mu_A(x_i))}-\sqrt{\mu_B(x_i)}\right)^{2(t+1)}}{\left(\sqrt{(1-\mu_A(x_i))\mu_B(x_i)}\right)^{t}}+\frac{\left(\sqrt{\mu_A(x_i)}-\sqrt{(1-\mu_B(x_i))}\right)^{2(t+1)}}{\left(\sqrt{\mu_A(x_i)(1-\mu_B(x_i))}\right)^{t}}\right]\\
&= h_t(A,\overline{B}).
\end{aligned}
$$

Hence, 6.4(c) holds.

6.4(d) It obviously follows from 6.4(b) and 6.4(c).

6.3 Applications of Generalized Hellinger's Fuzzy Divergence Measure in Practice

Fuzzy sets (FSs) are a suitable tool to handle with imprecisely defined facts and data, as well as with the vague knowledge. Next we provide the applications of the generalized Hellinger's fuzzy divergence measure in multi-criteria decision-making and medical diagnosis problems. Thereafter we present the application of the generalized Hellinger's fuzzy divergence in the context of multi-criteria decision-making.

6.3.1 Multi-criteria Decision-Making

Multi-criteria decision-making approach is considered as a key part of modern decision science and operational study. In practice MCDM problems are frequently experienced. For this we provide a method to solve multi-criteria decision-making problems with the help of the proposed generalized Hellinger's fuzzy divergence measure (6.2).

Let $A = \{A_1, A_2, \ldots, A_m\}$ be a set of options or alternatives, $C = \{C_1, C_2, \ldots, C_n\}$ be the set of criteria. The characteristics of the option A_i in terms of criteria C are represented by the following FSs:

$$A_i = \{\langle C_j, \mu_{ij}\rangle, C_j \in C\}, \quad i = 1, 2, 3, \ldots, m \text{ and } j = 1, 2, 3, \ldots, n$$

where μ_{ij} indicates the degree that the option A_i satisfies the criterion C_j.

We introduce the approach to solve the above multi-criteria fuzzy decision-making problem using Eq. (6.2). The computational procedure of the same is in order.

Step 1: Construct a fuzzy decision matrix

	C_1	C_2	C_3	...	C_n
A_1	d_{11}	d_{12}	d_{13}	...	d_{1n}
A_2	d_{21}	d_{22}	d_{23}	...	d_{2n}
A_3	d_{31}	d_{32}	d_{33}	...	d_{3n}
$\vdots$	$\vdots$	$\vdots$	$\vdots$	$\vdots$	$\vdots$
A_m	d_{m1}	d_{m2}	d_{m3}	...	d_{mn}

Step 2: Construct the normalized fuzzy decision matrix
This step converts the various dimensional criteria into non-dimensional criteria. In general, all of the criteria may be of the same type or different types. If the criteria are of different types, it is required to make them of the same type. For example, it is assumed that there are two types of criteria, say (i) benefit type, and (ii) cost type. Depending on the nature of criteria, we convert the cost criteria into the benefit criteria. For this, we here transform the fuzzy decision matrix $D = (d_{ij})_{m\times n}$ into the normalized fuzzy decision matrix say $R = (r_{ij})_{m\times n}$. An element r_{ij} of the normalized fuzzy decision matrix R is obtained as follows:

$$r_{ij} = \begin{cases} d_{ij}, \text{for benefit criteria } C_j \\ d_{ij}^c, \text{for cost criteria } C_j \end{cases}; \quad \text{where } i = 1, 2, \ldots, m;\ j = 1, 2, \ldots, n. \tag{6.5}$$

and d_{ij}^c is the complement of d_{ij} with $d_{ij}^c = 1 - \mu_{ij}$.

Step 3: Find out the positive-ideal solution A^+ and negative-ideal solution A^-

$$A^+ = \{\langle\mu_1^+\rangle, \langle\mu_2^+\rangle, \ldots, \langle\mu_n^+\rangle\} \quad \text{and} \quad A^- = \{\langle\mu_1^-\rangle, \langle\mu_2^-\rangle, \ldots, \langle\mu_n^-\rangle\}$$

where, for each $j = 1, 2, 3, \ldots, n$,

$$\left.\begin{aligned} \left\langle \mu_j^+ \right\rangle &= \left\langle \max_i \mu_{ij} \right\rangle \\ \left\langle \mu_j^- \right\rangle &= \left\langle \min_i \mu_{ij} \right\rangle \end{aligned}\right\} \tag{6.6}$$

Step 4: Calculate $h_t(A^+, A_i)$ and $h_t(A^-, A_i)$ using Eq. (6.2)
Step 5: Calculate the relative fuzzy divergence measure $h_t(A_i)$ of alternative A_i with respect to A^+ and A^-, where

$$h_t(A_i) = \frac{h_t(A^+, A_i)}{h_t(A^+, A_i) + h_t(A^-, A_i)}, \quad i = 1, 2, 3, \ldots, m \tag{6.7}$$

Step 6: Rank the preference order of all alternatives according to the relative fuzzy divergence measure.
Step 7: Select the best alternative A_k with smallest $h_t(A_k)$.

Numerical Example 6.1
We here demonstrate the applicability of the proposed generalized fuzzy divergence measure to solve a problem related to real multi-criteria decision-making. For this, we consider the decision-making problem of a customer while purchasing an air conditioner.

Example A customer wants to buy an air conditioner. There are four options (air condition systems) $A_i (i = 1, 2, 3, 4)$ available in the market. Suppose that three criteria: (i) C_1, economical; (ii) C_2, functional; and (iii) C_3, being operative, are taken into consideration during the selection procedure.

The four possible options (alternatives) are to be estimated by the customer under the above three criteria in the form of following sets:

$$\begin{aligned} A_1 &= \{\langle C_1, 0.2\rangle, \langle C_2, 0.4\rangle, \langle C_3, 0.6\rangle\}, \\ A_2 &= \{\langle C_1, 0.5\rangle, \langle C_2, 0.8\rangle, \langle C_3, 0.3\rangle\}, \\ A_3 &= \{\langle C_1, 0.7\rangle, \langle C_2, 0.3\rangle, \langle C_3, 0.6\rangle\}, \\ A_4 &= \{\langle C_1, 0.8\rangle, \langle C_2, 0.2\rangle, \langle C_3, 0.4\rangle\}. \end{aligned}$$

The step-wise computational procedure to solve the above multi-criteria fuzzy decision-making problem now goes as follows:

Step 1: Using above four options, a fuzzy decision matrix is constructed and presented in Table 6.1.

Table 6.1 A fuzzy decision matrix $D = (d_{ij})_{4\times3}$

	C_1	C_2	C_3
A_1	0.2	0.4	0.6
A_2	0.5	0.8	0.3
A_3	0.7	0.3	0.6
A_4	0.8	0.2	0.4

Table 6.2 The normalized fuzzy decision matrix $R = (r_{ij})_{4\times3}$

	C_1^c	C_2	C_3
A_1	0.8	0.4	0.6
A_2	0.5	0.8	0.3
A_3	0.3	0.3	0.6
A_4	0.2	0.2	0.4

Table 6.3 Calculated numerical values of $h_t(A^+, A_i)$, $t \in N$

	$t=1$	$t=2$	$t=5$
$h_t(A^+, A_1)$	0.8614	3.3229	136.0257
$h_t(A^+, A_2)$	0.9383	3.5941	158.7348
$h_t(A^+, A_3)$	2.6029	9.5052	299.2920
$h_t(A^+, A_4)$	4.1650	15.5051	477.3420

Step 2: Here C_1 is the cost criteria while other two are benefit criteria. We transform the cost criteria C_1 into benefit criteria C_1^c; we get the normalized fuzzy decision matrix in Table 6.2.
Step 3: The positive-ideal solution A^+ and negative-ideal solution A^-, respectively are

$$A^+ = \{\langle C_1, 0.8\rangle, \langle C_2, 0.8\rangle, \langle C_3, 0.6\rangle\},$$
$$A^- = \{\langle C_1, 0.2\rangle, \langle C_2, 0.2\rangle, \langle C_3, 0.3\rangle\}.$$

Step 4: Tables 6.3 and 6.4 show the calculated numerical values of $h_t(A^+, A_i)$ and $h_t(A^-, A_i)$ using Eq. (6.2) for $t \in N$.
Step 5: Calculated values of the relative fuzzy divergence measure $h_t(A_i)$ for $i = 1, 2, 3, 4$ with $t \in N$ are presented in Table 6.5.
Step 6: According to the calculated numerical values of relative divergences for different values of $t \in N$, ranking order of alternatives is given in Table 6.6.

Table 6.4 Calculated numerical values of $h_t(A^-, A_i)$, $t \in N$

	$t=1$	$t=2$	$t=5$
$h_t(A^-, A_1)$	2.6843	10.1811	377.1503
$h_t(A^-, A_2)$	2.5381	9.7895	364.6121
$h_t(A^-, A_3)$	0.5830	2.3506	97.3188
$h_t(A^-, A_4)$	0.0533	0.2070	8.8380

Table 6.5 Calculated numerical values of $h_t(A_i)$, $t \in N$

	$t = 1$	$t = 2$	$t = 5$
$h_t(A_1)$	0.2429	0.2461	0.2651
$h_t(A_2)$	0.2699	0.2685	0.3033
$h_t(A_3)$	0.8170	0.8017	0.7546
$h_t(A_4)$	0.9874	0.9868	0.9818

Table 6.6 Ranking order of alternatives $A_i (i = 1, 2, 3, 4)$, $t \in N$

	$t = 1$	$t = 2$	$t = 5$
A_1	1	1	1
A_2	2	2	2
A_3	3	3	3
A_4	4	4	4

Step 7: Thus we here find that variation in the values of t does not bring change in ranking. So A_1 is the most preferable alternative.

6.3.2 *Medical Diagnosis*

Many researchers have found the application of fuzzy sets theory in the diagnosis of diseases. Fuzzy pattern recognition algorithms are suitable tool to solve the medical diagnosis problems for recognizing the disease which is a challenging research area from a practical point of view. In order to illustrate the efficiency of the proposed generalized fuzzy divergence measure in solving medical diagnosis problems, using the algorithm given in Ohlan [8], we provide a numerical example.

Numerical Example 6.2
Suppose that the universe of discourse X is a set of symptoms

$$X = \{x_1(\text{Temperature}), x_2(\text{Headache}), x_3(\text{Stomach pain}), x_4(\text{Cough}), x_5(\text{Chest pain})\}$$

Consider a set of diagnosis

$$Q = \{Q_1(\text{Viral}), Q_2(\text{Malaria}), Q_3(\text{Typhoid}), Q_4(\text{Stomach problem}), Q_5(\text{Chest problem})\}$$

whose elements are presented by the following FSs, respectively,

$$\begin{aligned}
Q_1 &= \{\langle x_1, 0.7\rangle, \langle x_2, 0.2\rangle, \langle x_3, 0.0\rangle, \langle x_4, 0.7\rangle, \langle x_5, 0.1\rangle\},\\
Q_2 &= \{\langle x_1, 0.4\rangle, \langle x_2, 0.3\rangle, \langle x_3, 0.1\rangle, \langle x_4, 0.4\rangle, \langle x_5, 0.1\rangle\},\\
Q_3 &= \{\langle x_1, 0.1\rangle, \langle x_2, 0.2\rangle, \langle x_3, 0.8\rangle, \langle x_4, 0.2\rangle, \langle x_5, 0.2\rangle\},\\
Q_4 &= \{\langle x_1, 0.1\rangle, \langle x_2, 0.0\rangle, \langle x_3, 0.2\rangle, \langle x_4, 0.2\rangle, \langle x_5, 0.8\rangle\},\\
Q_5 &= \{\langle x_1, 0.3\rangle, \langle x_2, 0.6\rangle, \langle x_3, 0.2\rangle, \langle x_4, 0.2\rangle, \langle x_5, 0.1\rangle\}.
\end{aligned}$$

Table 6.7 The values of $h_t(P, Q_i)$, $i = \{1, 2, 3, 4, 5\}$ for $t \in N$

	$t = 1$	$t = 2$	$t = 5$	$t = 10$
$h_t(P, Q_1)$	**0.1863**	**0.8591**	**80.0827**	**1.4630e+005**
$h_t(P, Q_2)$	3.1666	15.6894	1.6507e+003	3.1634e+006
$h_t(P, Q_3)$	4.9445	24.0133	3.2880e+003	6.0278e+007
$h_t(P, Q_4)$	9.3597	42.3913	2.8381e+003	2.4434e+006
$h_t(P, Q_5)$	3.8857	14.7213	557.7232	1.7550e+005

The aim here is to assign a patient $P = \{\langle x_1, 0.8\rangle, \langle x_2, 0.2\rangle, \langle x_3, 0.6\rangle, \langle x_4, 0.6\rangle, \langle x_5, 0.1\rangle\}$, to one of the above-mentioned diagnosis Q_1, Q_2, Q_3, Q_4 and Q_5.

We proceed by considering the criteria $\min_{1 \le i \le 5}\{h_t(P, Q_i)\}$ with $t \in N$

Table 6.7 presents the values of $h_t(P, Q_i)$, $i = \{1, 2, 3, 4, 5\}$ for $t \in N$ using measure (6.2). It has been observed that the proper diagnosis for patient P is Q_1 (viral fever).

6.4 Concluding Remarks

In this chapter, we have presented generalized measure of discrimination in fuzzy environment. To do so, we have proposed and validated the generalized Hellinger's measure of fuzzy divergence. Particular cases and some of the interesting efficient properties of this divergence measure are proven. In this process, we provided a method for solving the problems related to multi-criteria decision-making and medical diagnosis in a fuzzy context with the illustrative example for each. We note that results from the method of multi-criteria decision-making present better alternative and an easier approach to decision-makers for evaluating issues which cannot be precisely defined. Thus, it is concluded that the proposed formula of generalized Hellinger's divergence measure and method of multi-criteria decision-making require no complicated computation. Finally, we observe that the presence of the parameters in the proposed measure provides a greater flexibility in applications.

References

2. Bellman RE, Zadeh LA (1970) Decision-making in a fuzzy environment. Manage Sci 17:141–164
1. Bhandari D, Pal NR (1993) Some new information measures for fuzzy sets. Inf Sci 67 (3):209–228
3. Duckstein L, Opricovic S (1980) Multiobjective optimization in river basin development. Water Resour Res 16(1):14–20

4. Dyer JS, Fishburn PC, Steuer RE, Wallenius J, Zionts S (1992) Multiple criteria decision making, multiattribute utility theory: the next ten years. Manage Sci 38(5):645–654
5. Erginel N, Çakmak T, Şentürk S (2010) Determining the preference of GSM operators in turkey with fuzzy topsis after mobile number portability system application. Anadolu Univ J Sci Tech Appl Sci Eng 11(2):81–93
6. Hellinger E (1909) Neue Begr¨undung der theorie der quadratischen formen von unendlichen vielen ver¨anderlichen. Journal für die reine und angewandte Mathematik 136:210–271
7. Hwang CL, Yoon K (1981) Multiple attribute decision making—methods and applications. Springer, New York
8. Ohlan A (2016) Intuitionistic fuzzy exponential divergence: application in multi-attribute decision making. J Intell Fuzzy Syst 30:1519–1530
9. Singh RP, Tomar VP (2008) On fuzzy divergence measures and their inequalities. Proceedings of 10th National Conference of ISITA 41–43
10. Taneja IJ (2013) Seven means, generalized triangular discrimination and generating divergence measures. Information 4:198–239
11. Zeleny M (1973) Compromise programming. In: Cochrane JL, Zeleny M (eds) Multiple criteria decision making. University of South Carolina Press, Columbia, SC, pp 262–301

Chapter 7
Intuitionistic Fuzzy Exponential Divergence and Multi-attribute Decision-Making

In this chapter, we apply the exponential approach on intuitionistic fuzzy sets (IFSs) and propose a new information-theoretic divergence measure, called intuitionistic fuzzy exponential divergence, to compute the difference between two IFSs.

Indeed, the notion of Atanassov's IFSs was first originated by Atanassov [1] which found to be well suited to deal with both fuzziness and lack of knowledge or non-specificity. It is noted that the concept of an IFS is the best alternative approach to define a fuzzy set (FS) in cases where existing information is not enough for the definition of imprecise concepts by means of a conventional FS. Thus, the concept of Atanassov IFSs is the generalization of the concept of FSs. Gau and Buehrer [9] introduced the notion of vague sets. But, Bustince and Burillo [6] presented that the notion of vague sets was equivalent to that of Atanassov IFSs. As a very significant content in fuzzy mathematics, the study on the divergence measure between IFSs has received more attention in recent years. Divergence measures of IFSs have been widely applied to many fields such as pattern recognition [10, 12, 18, 24], linguistic variables [11], medical diagnosis [7, 27], logical reasoning [14] and decision-making [17, 21, 22, 28]. Since the divergence measures of IFSs have been applied to many real-world situations, it is expected to have an efficient divergence measure which deals with the aspect of uncertainty, i.e. fuzziness and non-specificity or lack of knowledge.

Section 7.1 is devoted to review some relevant concepts related to fuzzy set theory and intuitionistic fuzzy set theory. In Sect. 7.2 we introduce the intuitionistic exponential divergence measure between IFSs and prove its essential properties. Some of elegant properties are proved on the proposed divergence measure in Sect. 7.3. In what follows, Sect. 7.4 presents the efficiency of the proposed intuitionistic fuzzy exponential divergence in pattern recognition by comparing it with some existing measures with the help of a numerical example. In this way, a method is presented to solve MADM problem using the proposed divergence measure in the intuitionistic fuzzy environment in Sect. 7.5. Section 7.6 presents

A. Ohlan and R. Ohlan, *Generalizations of Fuzzy Information Measures*,
DOI 10.1007/978-3-319-45928-8_7

the application of the proposed measure of intuitionistic fuzzy exponential divergence in the existing TOPSIS and MOORA methods of multi-attribute decision-making. A comparative analysis of the proposed method of multi-attribute decision-making and the existing methods is provided in the same section. The final section contains the concluding remarks of the work done in this chapter.

7.1 Preliminaries

We begin by reviewing some relevant concepts related to fuzzy set theory and intuitionistic fuzzy set theory.

Definition 7.1 (*Fuzzy Set [26]*) A fuzzy set A' defined on a finite universe of discourse $X = \{x_1, x_2, \ldots, x_n\}$ is given as

$$A' = \{\langle x_i, \mu_{A'}(x_i)\rangle / x_i \in X\} \tag{7.1}$$

where $\mu_{A'} : X \to [0, 1]$ is the membership function of A'. The membership value $\mu_{A'}(x_i)$ describes the degree of the belongingness of $x_i \in X$ in A'. When $\mu_{A'}(x_i)$ is valued in $\{0, 1\}$, it is the characteristic function of a crisp, i.e. non-fuzzy set.

Atanassov [1–4] introduced the concept of IFSs as the generalization of the concept of FSs.

Definition 7.2 (*Intuitionistic Fuzzy Set*) An intuitionistic fuzzy set A defined on a universe of discourse $X = \{x_1, x_2, \ldots, x_n\}$ is given as

$$A = \{\langle x_i, \mu_A(x_i), \nu_A(x_i)\rangle / x_i \in X\} \tag{7.2}$$

where $\mu_A : X \to [0, 1]$, $\nu_A : X \to [0, 1]$ with the condition $0 \leq \mu_A + \nu_A \leq 1 \;\; \forall x_i \in X$.

The numbers $\mu_A(x_i), \nu_A(x_i) \in [0, 1]$ denote the degree of membership and non-membership of x_i to A, respectively.

For each intuitionistic fuzzy set in X we will call $\pi_A(x_i) = 1 - \mu_A(x_i) - \nu_A(x_i)$, the intuitionistic index or degree of hesitation of x_i in A. It is obvious that $0 \leq \pi_A(x_i) \leq 1$ for each $x_i \in X$. For a fuzzy set A' in X, $\pi_A(x_i) = 0$ when $\nu_A(x_i) = 1 - \mu_A(x_i)$. Thus, FSs are the special cases of IFSs.

Atanassov [2] further defined set operations on intuitionistic fuzzy sets as follows:

Let $A, B \in IFS(X)$ given by

$$A = \{\langle x_i, \mu_A(x_i), \nu_A(x_i)\rangle / x_i \in X\},$$
$$B = \{\langle x_i, \mu_B(x_i), \nu_B(x_i)\rangle / x_i \in X\},$$

(i) $A \subseteq B$ iff $\mu_A(x_i) \le \mu_B(x_i)$ and $\nu_A(x_i) \ge \nu_B(x_i)$ $\forall x_i \in X$.
(ii) $A = B$ iff $A \subseteq B$ and $B \subseteq A$.
(iii) $A^c = \{\langle x_i, \nu_A(x_i), \mu_A(x_i)\rangle / x_i \in X\}$.
(iv) $A \cup B = \{\langle x_i, \max(\mu_A(x_i), \mu_B(x_i)), \min(\nu_A(x_i), \nu_B(x_i))\rangle / x_i \in X\}$.
(v) $A \cap B = \{\langle x_i, \min(\mu_A(x_i), \mu_B(x_i)), \max(\nu_A(x_i), \nu_B(x_i))\rangle / x_i \in X\}$.

7.2 Definition of Intuitionistic Fuzzy Exponential Divergence Measure

In this section, the intuitionistic fuzzy exponential divergence between IFSs is proposed and some of its properties are proved. Li et al. [15] introduced a method for transforming Atanassov IFSs into FSs as briefly mentioned in Verma and Sharma [23] that was used to propose the intuitionistic exponential divergence measure between IFSs.

$$I_1(A,B) = \sum_{i=1}^{n}\left[\frac{1}{2} - \left(1 - \frac{\mu_A(x_i) + 1 - \nu_A(x_i)}{2}\right)e^{\left(\frac{\mu_A(x_i)+1-\nu_A(x_i)}{2} - \frac{\mu_B(x_i)+1-\nu_B(x_i)}{2}\right)}\right] \tag{7.3}$$

Similarly, the expected amount of information for discrimination of A^c against B^c is given by

$$I_1(A^c,B^c) = \sum_{i=1}^{n}\left[\frac{1}{2} - \left(\frac{\mu_A(x_i) + 1 - \nu_A(x_i)}{2}\right)e^{\left(\frac{\mu_B(x_i)+1-\nu_B(x_i)}{2} - \frac{\mu_A(x_i)+1-\nu_A(x_i)}{2}\right)}\right] \tag{7.4}$$

Now, $I_1(A,B) \neq I_1(A^c,B^c)$. Thus we propose the discrimination measure of information between IFSs A and B given by

$$I(A,B) = \sum_{i=1}^{n}\left[\begin{array}{l} 1 - \left(\frac{\nu_A(x_i)+1-\mu_A(x_i)}{2}\right)e^{\left(\frac{((\mu_A(x_i)-\mu_B(x_i))-(\nu_A(x_i)-\nu_B(x_i)))}{2}\right)} \\ -\left(\frac{\mu_A(x_i)+1-\nu_A(x_i)}{2}\right)e^{\left(\frac{((\mu_B(x_i)-\mu_A(x_i))-(\nu_B(x_i)-\nu_A(x_i)))}{2}\right)} \end{array}\right] \tag{7.5}$$

Similarly, we have

$$I(B,A) = \sum_{i=1}^{n}\left[\begin{array}{l} 1 - \left(\frac{\nu_B(x_i)+1-\mu_B(x_i)}{2}\right)e^{\left(\frac{((\mu_B(x_i)-\mu_A(x_i))-(\nu_B(x_i)-\nu_A(x_i)))}{2}\right)} \\ -\left(\frac{\mu_B(x_i)+1-\nu_B(x_i)}{2}\right)e^{\left(\frac{((\mu_A(x_i)-\mu_B(x_i))-(\nu_A(x_i)-\nu_B(x_i)))}{2}\right)} \end{array}\right] \tag{7.6}$$

Thus, we propose the intuitionistic fuzzy exponential divergence between IFSs A and B of universe of discourse $X = \{x_1, x_2, \ldots, x_n\}$ is given by

$$D^E_{IFS}(A,B) = I(A,B) + I(B,A)$$

$$= \sum_{i=1}^{n} \left[\begin{array}{l} 2 - \left(\frac{\nu_A(x_i) + 1 - \mu_A(x_i)}{2} + \frac{\mu_B(x_i) + 1 - \nu_B(x_i)}{2} \right) e^{\left(\frac{((\mu_A(x_i) - \mu_B(x_i)) - (\nu_A(x_i) - \nu_B(x_i)))}{2} \right)} \\ - \left(\frac{\mu_A(x_i) + 1 - \nu_A(x_i)}{2} + \frac{\nu_B(x_i) + 1 - \mu_B(x_i)}{2} \right) e^{\left(\frac{((\mu_B(x_i) - \mu_A(x_i)) - (\nu_B(x_i) - \nu_A(x_i)))}{2} \right)} \end{array} \right]$$

$$D^E_{IFS}(A,B) = \sum_{i=1}^{n} \left[\begin{array}{l} 2 - \left(1 - \frac{(\mu_A(x_i) - \mu_B(x_i)) - (\nu_A(x_i) - \nu_B(x_i))}{2} \right) e^{\left(\frac{(\mu_A(x_i) - \mu_B(x_i)) - (\nu_A(x_i) - \nu_B(x_i))}{2} \right)} \\ - \left(1 + \frac{(\mu_A(x_i) - \mu_B(x_i)) - (\nu_A(x_i) - \nu_B(x_i))}{2} \right) e^{\left(\frac{(\nu_A(x_i) - \nu_B(x_i)) - (\mu_A(x_i) - \mu_B(x_i))}{2} \right)} \end{array} \right] \tag{7.7}$$

It is natural to then ask 'is the defined intuitionistic divergence measure $D^E_{IFS}(A,B)$ reasonable?' We answer this question in the following theorem satisfying the properties defined in [11].

Theorem 7.1 *The defined fuzzy exponential divergence measure* $D^E_{IFS}(A,B)$ *in (7.7) between IFSs A and B satisfies the following properties:*

(i) $0 \le D^E_{IFS}(A,B) \le 1$
(ii) $D^E_{IFS}(A,B) = 0$ if and only if $A = B$.
(iii) $D^E_{IFS}(A,B) = D^E_{IFS}(B,A)$
(iv) If $A \subseteq B \subseteq C$, $A, B, C \in IFSs(X)$

Then $D^E_{IFS}(A,B) \le D^E_{IFS}(A,C)$ and $D^E_{IFS}(B,C) \le D^E_{IFS}(A,C)$.

Proof (i) Since (7.7) is a convex function, refer Theorem 3.1 of [8]. Then $D^E_{IFS}(A,B)$ increases as $\|A - B\|_1$ increases

Where $\|A - B\|_1 = |\mu_A(x_i) - \mu_B(x_i)| + |\nu_A(x_i) - \nu_B(x_i)|$.

Then $D^E_{IFS}(A,B)$ attains its maximum value at the following degenerate cases:

$$A = (1,0), B = (0,1) \quad \text{or} \quad A = (0,1), B = (1,0).$$

It gives us that $0 \le D^E_{IFS}(A,B) \le 1$.

Obviously, the properties (ii) and (iii) are satisfied by $D^E_{IFS}(A,B)$.

(iv) For $A \subseteq B \subseteq C$ we have $\|A - B\|_1 \le \|A - C\|_1$ and $\|B - C\|_1 \le \|A - C\|_1$

Thus $D^E_{IFS}(A,B) \le D^E_{IFS}(A,C)$ and $D^E_{IFS}(B,C) \le D^E_{IFS}(A,C)$.

Hence in view of definition of [11], $D^E_{IFS}(A,B)$ is a valid measure of divergence between IFSs A and B.

7.3 Properties of Intuitionistic Fuzzy Exponential Divergence Measure

In this section, we provide additional properties of the proposed intuitionistic fuzzy exponential divergence measure (7.7) in the following theorems. While proving these theorems we consider the separation of X into two parts X_1 and X_2, such that

$$X_1 = \{x_i/x_i \in X, A(x_i) \subseteq B(x_i)\} \text{ and}$$
$$X_2 = \{x_i/x_i \in X, A(x_i) \supseteq B(x_i)\}$$

where $A(x_i) = \{(x_i, \mu_A(x_i), \nu_A(x_i))\}$ and $B(x_i) = \{(x_i, \mu_B(x_i), \nu_B(x_i))\}$.

Thus on using the operations explained above in Sect. 7.2, we get

In set X_1, $\mu_A(x_i) \leq \mu_B(x_i)$ and $\nu_A(x_i) \geq \nu_B(x_i)$.

In set X_2, $\mu_A(x_i) \geq \mu_B(x_i)$ and $\nu_A(x_i) \leq \nu_B(x_i)$.

Theorem 7.2 *For $A, B, C \in IFS(X)$,*

(a) $D^E_{IFS}(A \cup B, C) \leq D^E_{IFS}(A, C) + D^E_{IFS}(B, C)$.

(b) $D^E_{IFS}(A \cap B, C) \leq D^E_{IFS}(A, C) + D^E_{IFS}(B, C)$.

Proof 7.2(a) Let us consider the expression

$$\begin{aligned}
&D^E_{IFS}(A, C) + D^E_{IFS}(B, C) - D^E_{IFS}(A \cup B, C)\\
&= \sum_{i=1}^{n} \left[\begin{array}{l} 2 - \left(1 - \frac{(\mu_A(x_i) - \mu_C(x_i)) - (\nu_A(x_i) - \nu_C(x_i))}{2}\right) e^{\left(\frac{(\mu_A(x_i) - \mu_C(x_i)) - (\nu_A(x_i) - \nu_C(x_i))}{2}\right)} \\ - \left(1 + \frac{(\mu_A(x_i) - \mu_C(x_i)) - (\nu_A(x_i) - \nu_C(x_i))}{2}\right) e^{\left(\frac{(\nu_A(x_i) - \nu_C(x_i)) - (\mu_A(x_i) - \mu_C(x_i))}{2}\right)} \end{array} \right]\\
&+ \sum_{i=1}^{n} \left[\begin{array}{l} 2 - \left(1 - \frac{(\mu_B(x_i) - \mu_C(x_i)) - (\nu_B(x_i) - \nu_C(x_i))}{2}\right) e^{\left(\frac{(\mu_B(x_i) - \mu_C(x_i)) - (\nu_B(x_i) - \nu_C(x_i))}{2}\right)} \\ - \left(1 + \frac{(\mu_B(x_i) - \mu_C(x_i)) - (\nu_B(x_i) - \nu_C(x_i))}{2}\right) e^{\left(\frac{(\nu_B(x_i) - \nu_C(x_i)) - (\mu_B(x_i) - \mu_C(x_i))}{2}\right)} \end{array} \right]\\
&- \sum_{i=1}^{n} \left[\begin{array}{l} 2 - \left(1 - \frac{(\mu_{A \cup B}(x_i) - \mu_C(x_i)) - (\nu_{A \cup B}(x_i) - \nu_C(x_i))}{2}\right) e^{\left(\frac{(\mu_{A \cup B}(x_i) - \mu_C(x_i)) - (\nu_{A \cup B}(x_i) - \nu_C(x_i))}{2}\right)} \\ - \left(1 + \frac{(\mu_{A \cup B}(x_i) - \mu_C(x_i)) - (\nu_{A \cup B}(x_i) - \nu_C(x_i))}{2}\right) e^{\left(\frac{(\nu_{A \cup B}(x_i) - \nu_C(x_i)) - (\mu_{A \cup B}(x_i) - \mu_C(x_i))}{2}\right)} \end{array} \right]\\
&= \sum_{X_1} \left[\begin{array}{l} 2 - \left(1 - \frac{(\mu_A(x_i) - \mu_C(x_i)) - (\nu_A(x_i) - \nu_C(x_i))}{2}\right) e^{\left(\frac{(\mu_A(x_i) - \mu_C(x_i)) - (\nu_A(x_i) - \nu_C(x_i))}{2}\right)} \\ - \left(1 + \frac{(\mu_A(x_i) - \mu_C(x_i)) - (\nu_A(x_i) - \nu_C(x_i))}{2}\right) e^{\left(\frac{(\nu_A(x_i) - \nu_C(x_i)) - (\mu_A(x_i) - \mu_C(x_i))}{2}\right)} \end{array} \right]\\
&+ \sum_{X_2} \left[\begin{array}{l} 2 - \left(1 - \frac{(\mu_B(x_i) - \mu_C(x_i)) - (\nu_B(x_i) - \nu_C(x_i))}{2}\right) e^{\left(\frac{(\mu_B(x_i) - \mu_C(x_i)) - (\nu_B(x_i) - \nu_C(x_i))}{2}\right)} \\ - \left(1 + \frac{(\mu_B(x_i) - \mu_C(x_i)) - (\nu_B(x_i) - \nu_C(x_i))}{2}\right) e^{\left(\frac{(\nu_B(x_i) - \nu_C(x_i)) - (\mu_B(x_i) - \mu_C(x_i))}{2}\right)} \end{array} \right]\\
&\geq 0.
\end{aligned}$$

Hence, 7.2(a) holds.
Similarly, 7.2(b) can also be proved.

Theorem 7.3 *For* $A, B, C \in IFS(X)$,

$$D_{IFS}^{E}(A \cup B, C) + D_{IFS}^{E}(A \cap B, C) = D_{IFS}^{E}(A, C) + D_{IFS}^{E}(B, C).$$

Proof $D_{IFS}^{E}(A \cup B, C)$

$$= \sum_{i=1}^{n} \left[\begin{array}{l} 2 - \left(1 - \frac{(\mu_{A \cup B}(x_i) - \mu_C(x_i)) - (\nu_{A \cup B}(x_i) - \nu_C(x_i))}{2}\right) e^{\left(\frac{(\mu_{A \cup B}(x_i) - \mu_C(x_i)) - (\nu_{A \cup B}(x_i) - \nu_C(x_i))}{2}\right)} \\ - \left(1 + \frac{(\mu_{A \cup B}(x_i) - \mu_C(x_i)) - (\nu_{A \cup B}(x_i) - \nu_C(x_i))}{2}\right) e^{\left(\frac{(\nu_{A \cup B}(x_i) - \nu_C(x_i)) - (\mu_{A \cup B}(x_i) - \mu_C(x_i))}{2}\right)} \end{array} \right]$$

$$= \sum_{X_1} \left[\begin{array}{l} 2 - \left(1 - \frac{(\mu_B(x_i) - \mu_C(x_i)) - (\nu_B(x_i) - \nu_C(x_i))}{2}\right) e^{\left(\frac{(\mu_B(x_i) - \mu_C(x_i)) - (\nu_B(x_i) - \nu_C(x_i))}{2}\right)} \\ - \left(1 + \frac{(\mu_B(x_i) - \mu_C(x_i)) - (\nu_B(x_i) - \nu_C(x_i))}{2}\right) e^{\left(\frac{(\nu_B(x_i) - \nu_C(x_i)) - (\mu_B(x_i) - \mu_C(x_i))}{2}\right)} \end{array} \right]$$
$$+ \sum_{X_2} \left[\begin{array}{l} 2 - \left(1 - \frac{(\mu_A(x_i) - \mu_C(x_i)) - (\nu_A(x_i) - \nu_C(x_i))}{2}\right) e^{\left(\frac{(\mu_A(x_i) - \mu_C(x_i)) - (\nu_A(x_i) - \nu_C(x_i))}{2}\right)} \\ - \left(1 + \frac{(\mu_A(x_i) - \mu_C(x_i)) - (\nu_A(x_i) - \nu_C(x_i))}{2}\right) e^{\left(\frac{(\nu_A(x_i) - \nu_C(x_i)) - (\mu_A(x_i) - \mu_C(x_i))}{2}\right)} \end{array} \right] \tag{7.8}$$

Now $D_{IFS}^{E}(A \cap B, C)$

$$= \sum_{i=1}^{n} \left[\begin{array}{l} 2 - \left(1 - \frac{(\mu_{A \cap B}(x_i) - \mu_C(x_i)) - (\nu_{A \cap B}(x_i) - \nu_C(x_i))}{2}\right) e^{\left(\frac{(\mu_{A \cap B}(x_i) - \mu_C(x_i)) - (\nu_{A \cap B}(x_i) - \nu_C(x_i))}{2}\right)} \\ - \left(1 + \frac{(\mu_{A \cap B}(x_i) - \mu_C(x_i)) - (\nu_{A \cap B}(x_i) - \nu_C(x_i))}{2}\right) e^{\left(\frac{(\nu_{A \cap B}(x_i) - \nu_C(x_i)) - (\mu_{A \cap B}(x_i) - \mu_C(x_i))}{2}\right)} \end{array} \right]$$

$$= \sum_{X_1} \left[\begin{array}{l} 2 - \left(1 - \frac{(\mu_A(x_i) - \mu_C(x_i)) - (\nu_A(x_i) - \nu_C(x_i))}{2}\right) e^{\left(\frac{(\mu_A(x_i) - \mu_C(x_i)) - (\nu_A(x_i) - \nu_C(x_i))}{2}\right)} \\ - \left(1 + \frac{(\mu_A(x_i) - \mu_C(x_i)) - (\nu_A(x_i) - \nu_C(x_i))}{2}\right) e^{\left(\frac{(\nu_A(x_i) - \nu_C(x_i)) - (\mu_A(x_i) - \mu_C(x_i))}{2}\right)} \end{array} \right]$$
$$+ \sum_{X_2} \left[\begin{array}{l} 2 - \left(1 - \frac{(\mu_B(x_i) - \mu_C(x_i)) - (\nu_B(x_i) - \nu_C(x_i))}{2}\right) e^{\left(\frac{(\mu_B(x_i) - \mu_C(x_i)) - (\nu_B(x_i) - \nu_C(x_i))}{2}\right)} \\ - \left(1 + \frac{(\mu_B(x_i) - \mu_C(x_i)) - (\nu_B(x_i) - \nu_C(x_i))}{2}\right) e^{\left(\frac{(\nu_B(x_i) - \nu_C(x_i)) - (\mu_B(x_i) - \mu_C(x_i))}{2}\right)} \end{array} \right] \tag{7.9}$$

On adding Eqs. (7.8) and (7.9) we get the result.
Hence, 7.3 holds.

Theorem 7.4 *For* $A, B, C \in IFS(X)$,

(a) $D^E_{IFS}(A \cup B, A \cap B) = D^E_{IFS}(A, B)$.
(b) $D^E_{IFS}(A \cap B, A \cup B) = D^E_{IFS}(A, B)$.

Proof 7.4(a) $D^E_{IFS}(A \cup B, A \cap B)$

$$= \sum_{i=1}^{n} \left[\begin{array}{l} 2 - \left(1 - \frac{(\mu_{A\cup B}(x_i) - \mu_{A\cap B}(x_i)) - (\nu_{A\cup B}(x_i) - \nu_{A\cap B}(x_i))}{2}\right) \\ e^{\left(\frac{(\mu_{A\cup B}(x_i) - \mu_{A\cap B}(x_i)) - (\nu_{A\cup B}(x_i) - \nu_{A\cap B}(x_i))}{2}\right)} \\ \quad - \left(1 + \frac{(\mu_{A\cup B}(x_i) - \mu_{A\cap B}(x_i)) - (\nu_{A\cup B}(x_i) - \nu_{A\cap B}(x_i))}{2}\right) \\ e^{\left(\frac{(\nu_{A\cup B}(x_i) - \nu_{A\cap B}(x_i)) - (\mu_{A\cup B}(x_i) - \mu_{A\cap B}(x_i))}{2}\right)} \end{array} \right]$$

$$= \sum_{X_1} \left[\begin{array}{l} 2 - \left(1 - \frac{(\mu_B(x_i) - \mu_A(x_i)) - (\nu_B(x_i) - \nu_A(x_i))}{2}\right) e^{\left(\frac{(\mu_B(x_i) - \mu_A(x_i)) - (\nu_B(x_i) - \nu_A(x_i))}{2}\right)} \\ - \left(1 + \frac{(\mu_B(x_i) - \mu_A(x_i)) - (\nu_B(x_i) - \nu_A(x_i))}{2}\right) e^{\left(\frac{(\nu_B(x_i) - \nu_A(x_i)) - (\mu_B(x_i) - \mu_A(x_i))}{2}\right)} \end{array} \right]$$

$$+ \sum_{X_2} \left[\begin{array}{l} 2 - \left(1 - \frac{(\mu_A(x_i) - \mu_B(x_i)) - (\nu_A(x_i) - \nu_B(x_i))}{2}\right) e^{\left(\frac{(\mu_A(x_i) - \mu_B(x_i)) - (\nu_A(x_i) - \nu_B(x_i))}{2}\right)} \\ - \left(1 + \frac{(\mu_A(x_i) - \mu_B(x_i)) - (\nu_A(x_i) - \nu_B(x_i))}{2}\right) e^{\left(\frac{(\nu_A(x_i) - \nu_B(x_i)) - (\mu_A(x_i) - \mu_B(x_i))}{2}\right)} \end{array} \right]$$

$$= D^E_{IFS}(A, B).$$

Hence, 7.4(a) holds.
Similarly, 7.4(b) can be proved.

Theorem 7.5 *For* $A, B, C \in IFS(X)$,

(a) $D^E_{IFS}(A, A \cup B) + D^E_{IFS}(A, A \cap B) = D^E_{IFS}(A, B)$.
(b) $D^E_{IFS}(B, A \cup B) + D^E_{IFS}(B, A \cap B) = D^E_{IFS}(A, B)$.

Proof 7.5(a)

$$D^E_{IFS}(A, A\cup B) = \sum_{i=1}^{n}\left[\begin{array}{l} 2-\left(1-\frac{(\mu_A(x_i)-\mu_{A\cup B}(x_i))-(\nu_A(x_i)-\nu_{A\cup B}(x_i))}{2}\right) \\ e^{\left(\frac{(\mu_A(x_i)-\mu_{A\cup B}(x_i))-(\nu_A(x_i)-\nu_{A\cup B}(x_i))}{2}\right)} \\ \quad -\left(1+\frac{(\mu_A(x_i)-\mu_{A\cup B}(x_i))-(\nu_A(x_i)-\nu_{A\cup B}(x_i))}{2}\right) \\ e^{\left(\frac{(\nu_A(x_i)-\nu_{A\cup B}(x_i))-(\mu_A(x_i)-\mu_{A\cup B}(x_i))}{2}\right)} \end{array}\right]$$

$$= \sum_{X_1}\left[\begin{array}{l} 2-\left(1-\frac{(\mu_A(x_i)-\mu_B(x_i))-(\nu_A(x_i)-\nu_B(x_i))}{2}\right)e^{\left(\frac{(\mu_A(x_i)-\mu_B(x_i))-(\nu_A(x_i)-\nu_B(x_i))}{2}\right)} \\ -\left(1+\frac{(\mu_A(x_i)-\mu_B(x_i))-(\nu_A(x_i)-\nu_B(x_i))}{2}\right)e^{\left(\frac{(\nu_A(x_i)-\nu_B(x_i))-(\mu_A(x_i)-\mu_B(x_i))}{2}\right)} \end{array}\right] \tag{7.10}$$

Now

$$D^E_{IFS}(A, A\cap B) = \sum_{i=1}^{n}\left[\begin{array}{l} 2-\left(1-\frac{(\mu_A(x_i)-\mu_{A\cap B}(x_i))-(\nu_A(x_i)-\nu_{A\cap B}(x_i))}{2}\right)e^{\left(\frac{(\mu_A(x_i)-\mu_{A\cap B}(x_i))-(\nu_A(x_i)-\nu_{A\cap B}(x_i))}{2}\right)} \\ -\left(1+\frac{(\mu_A(x_i)-\mu_{A\cap B}(x_i))-(\nu_A(x_i)-\nu_{A\cap B}(x_i))}{2}\right)e^{\left(\frac{(\nu_A(x_i)-\nu_{A\cap B}(x_i))-(\mu_A(x_i)-\mu_{A\cap B}(x_i))}{2}\right)} \end{array}\right]$$

$$= \sum_{X_2}\left[\begin{array}{l} 2-\left(1-\frac{(\mu_A(x_i)-\mu_B(x_i))-(\nu_A(x_i)-\nu_B(x_i))}{2}\right)e^{\left(\frac{(\mu_A(x_i)-\mu_B(x_i))-(\nu_A(x_i)-\nu_B(x_i))}{2}\right)} \\ -\left(1+\frac{(\mu_A(x_i)-\mu_B(x_i))-(\nu_A(x_i)-\nu_B(x_i))}{2}\right)e^{\left(\frac{(\nu_A(x_i)-\nu_B(x_i))-(\mu_A(x_i)-\mu_B(x_i))}{2}\right)} \end{array}\right] \tag{7.11}$$

On adding Eqs. (7.10) and (7.11), we get the result.
Hence, 7.5(a) holds.
Similarly, 7.5(b) can be proved.

Theorem 7.6 *For $A, B, C \in IFS(X)$,*

(a) $D^E_{IFS}(A, B) = D^E_{IFS}(A^c, B^c)$
(b) $D^E_{IFS}(A^c, B) = D^E_{IFS}(A, B^c)$
(c) $D^E_{IFS}(A, B) + D^E_{IFS}(A^c, B) = D^E_{IFS}(A^c, B^c) + D^E_{IFS}(A^c, B)$.

Proof 7.6(a) $D^E_{IFS}(A^c, B^c)$

$$
= \sum_{i=1}^{n} \left[\begin{array}{l} 2 - \left(1 - \frac{(\mu_{A^c}(x_i) - \mu_{B^c}(x_i)) - (\nu_{A^c}(x_i) - \nu_{B^c}(x_i))}{2}\right) e^{\left(\frac{(\mu_{A^c}(x_i)-\mu_{B^c}(x_i))-(\nu_{A^c}(x_i)-\nu_B(x_i))}{2}\right)} \\ -\left(1 + \frac{(\mu_{A^c}(x_i) - \mu_{B^c}(x_i)) - (\nu_{A^c}(x_i) - \nu_{B^c}(x_i))}{2}\right) e^{\left(\frac{(\nu_{A^c}(x_i)-\nu_{B^c}(x_i))-(\mu_{A^c}(x_i)-\mu_{B^c}(x_i))}{2}\right)} \end{array} \right]
$$

$$
= \sum_{i=1}^{n} \left[\begin{array}{l} 2 - \left(1 - \frac{(\nu_A(x_i) - \nu_B(x_i)) - (\mu_A(x_i) - \mu_B(x_i))}{2}\right) e^{\left(\frac{(\nu_A(x_i)-\nu_B(x_i))-(\mu_A(x_i)-\mu_B(x_i))}{2}\right)} \\ -\left(1 + \frac{(\nu_A(x_i) - \nu_B(x_i)) - (\mu_A(x_i) - \mu_B(x_i))}{2}\right) e^{\left(\frac{(\mu_A(x_i)-\mu_B(x_i))-(\nu_A(x_i)-\nu_B(x_i))}{2}\right)} \end{array} \right]
$$

$$
= \sum_{i=1}^{n} \left[\begin{array}{l} 2 - \left(1 + \frac{(\mu_A(x_i) - \mu_B(x_i)) - (\nu_A(x_i) - \nu_B(x_i))}{2}\right) e^{\left(\frac{(\nu_A(x_i)-\nu_B(x_i))-(\mu_A(x_i)-\mu_B(x_i))}{2}\right)} \\ -\left(1 - \frac{(\mu_A(x_i) - \mu_B(x_i)) - (\nu_A(x_i) - \nu_B(x_i))}{2}\right) e^{\left(\frac{(\mu_A(x_i)-\mu_B(x_i))-(\nu_A(x_i)-\nu_B(x_i))}{2}\right)} \end{array} \right]
$$

$$
= D_{IFS}^{E}(A, B).
$$

Hence, 7.6(a) holds.

7.6(b) $D_{IFS}^{E}(A^c, B)$

$$
= \sum_{i=1}^{n} \left[\begin{array}{l} 2 - \left(1 - \frac{(\mu_{A^c}(x_i) - \mu_B(x_i)) - (\nu_{A^c}(x_i) - \nu_B(x_i))}{2}\right) e^{\left(\frac{(\mu_{A^c}(x_i)-\mu_B(x_i))-(\nu_{A^c}(x_i)-\nu_B(x_i))}{2}\right)} \\ -\left(1 + \frac{(\mu_{A^c}(x_i) - \mu_B(x_i)) - (\nu_{A^c}(x_i) - \nu_B(x_i))}{2}\right) e^{\left(\frac{(\nu_{A^c}(x_i)-\nu_B(x_i))-(\mu_{A^c}(x_i)-\mu_B(x_i))}{2}\right)} \end{array} \right]
$$

$$
= \sum_{i=1}^{n} \left[\begin{array}{l} 2 - \left(1 - \frac{(\nu_A(x_i) - \mu_B(x_i)) - (\mu_A(x_i) - \nu_B(x_i))}{2}\right) e^{\left(\frac{(\nu_A(x_i)-\mu_B(x_i))-(\mu_A(x_i)-\nu_B(x_i))}{2}\right)} \\ -\left(1 + \frac{(\nu_A(x_i) - \mu_B(x_i)) - (\mu_A(x_i) - \nu_B(x_i))}{2}\right) e^{\left(\frac{(\mu_A(x_i)-\nu_B(x_i))-(\nu_A(x_i)-\mu_B(x_i))}{2}\right)} \end{array} \right]
$$

Now $D_{IFS}^{E}(A, B^c)$

$$
= \sum_{i=1}^{n} \left[\begin{array}{l} 2 - \left(1 - \frac{(\mu_A(x_i) - \mu_{B^c}(x_i)) - (\nu_A(x_i) - \nu_{B^c}(x_i))}{2}\right) e^{\left(\frac{(\mu_A(x_i)-\mu_{B^c}(x_i))-(\nu_A(x_i)-\nu_{B^c}(x_i))}{2}\right)} \\ -\left(1 + \frac{(\mu_A(x_i) - \mu_{B^c}(x_i)) - (\nu_A(x_i) - \nu_{B^c}(x_i))}{2}\right) e^{\left(\frac{(\nu_A(x_i)-\nu_{B^c}(x_i))-(\mu_A(x_i)-\mu_{B^c}(x_i))}{2}\right)} \end{array} \right]
$$

$$
= \sum_{i=1}^{n} \left[\begin{array}{l} 2 - \left(1 - \frac{(\mu_A(x_i) - \nu_B(x_i)) - (\nu_A(x_i) - \mu_B(x_i))}{2}\right) e^{\left(\frac{(\mu_A(x_i)-\nu_B(x_i))-(\nu_A(x_i)-\mu_B(x_i))}{2}\right)} \\ -\left(1 + \frac{(\mu_A(x_i) - \nu_B(x_i)) - (\nu_A(x_i) - \mu_B(x_i))}{2}\right) e^{\left(\frac{(\nu_A(x_i)-\mu_B(x_i))-(\mu_A(x_i)-\nu_B(x_i))}{2}\right)} \end{array} \right]
$$

$$
= \sum_{i=1}^{n} \left[\begin{array}{l} 2 - \left(1 - \frac{(\nu_A(x_i) - \mu_B(x_i)) - (\mu_A(x_i) - \nu_B(x_i))}{2}\right) e^{\left(\frac{(\nu_A(x_i)-\mu_B(x_i))-(\mu_A(x_i)-\nu_B(x_i))}{2}\right)} \\ -\left(1 + \frac{(\nu_A(x_i) - \mu_B(x_i)) - (\mu_A(x_i) - \nu_B(x_i))}{2}\right) e^{\left(\frac{(\mu_A(x_i)-\nu_B(x_i))-(\nu_A(x_i)-\mu_B(x_i))}{2}\right)} \end{array} \right]
$$

$$
= D_{IFS}^{E}(A^c, B).
$$

Hence, 7.6(b) holds.
7.6(c) It directly follows from 7.6(a) and 7.6(b).

7.4 Numerical Example

We now demonstrate the efficiency of the proposed intuitionistic fuzzy exponential divergence measure in the context of pattern recognition by comparing it with the existing intuitionistic fuzzy divergence measures presented in [16, 11, 24, 27].

Let A and B be two intuitionistic fuzzy sets in the universe of discourse $X = \{x_1, x_2, \ldots x_n\}$. Li [16] introduced the dissimilarity measure for intuitionistic fuzzy sets given by

$$d(A,B) = \sum_{x \in X} [|\mu_A(x) - \mu_B(x)| + |\nu_A(x) + \nu_B(x)|]/2. \tag{7.12}$$

Hung and Yang [11] defined the distance between intuitionistic fuzzy sets A and B using hausdorff distance as follows:

$$d_H(A,B) = \frac{1}{n} H(I_A(x_i), I_B(x_i)) \tag{7.13}$$

where $I_A(x_i)$ and $I_B(x_i)$ be subintervals on [0, 1] denoted by $I_A(x_i) = [\mu_A(x_i), 1 - \nu_A(x_i)]$ and $I_B(x_i) = [\mu_B(x_i), 1 - \nu_B(x_i)]$ and the Hausdorff distance $H(A,B)$ = max $\{|a_1 - b_1|, |a_2 - b_2|\}$ is defined for two intervals $A = [a_1, a_2]$ and $B = [b_1, b_2]$.

Vlachos and Sergiadis [24] provided the intuitionistic fuzzy divergence measure given by

$$D_{IFS}(A,B) = I_{IFS}(A,B) + I_{IFS}(B,A) \tag{7.14}$$

where

$$I_{IFS}(A,B) = \sum_{i=1}^{n} \left[\mu_A(x_i) In \frac{\mu_A(x_i)}{\frac{1}{2}(\mu_A(x_i) + \mu_B(x_i))} + \nu_A(x_i) In \frac{\nu_A(x_i)}{\frac{1}{2}(\nu_A(x_i) + \nu_B(x_i))} \right]$$

where $I^{\mu}(A,B) = In \frac{\mu_A(x_i)}{\mu_B(x_i)}$ is the amount of discrimination of $\mu_A(x_i)$ from $\mu_B(x_i)$.

Zhang and Jiang [27] presented a measure of divergence between IFSs/vague sets A and B as

$$D^*(A,B) = D(A,B) + D(B,A) \tag{7.15}$$

where

$$D(A,B) = \sum_{i=1}^{n} \left[\begin{array}{l} \left(\frac{\mu_A(x_i)+1-\nu_A(x_i)}{2}\right)\log_2 \frac{[\mu_A(x_i)+1-\nu_A(x_i)]/2}{\frac{1}{4}\{[\mu_A(x_i)+1-\nu_A(x_i)]+[\mu_B(x_i)+1-\nu_B(x_i)]\}} \\ +\left(\frac{1-\mu_A(x_i)+\nu_A(x_i)}{2}\right)\log_2 \frac{[1-\mu_A(x_i)+\nu_A(x_i)]/2}{\frac{1}{4}\{[1-\mu_A(x_i)+\nu_A(x_i)]+[1-\mu_B(x_i)+\nu_B(x_i)]\}} \end{array} \right].$$

Here we use the following procedure to solve pattern recognition problem in intuitionistic fuzzy environment.

Suppose that we are given m known patterns $P_1, P_2, P_3, \ldots, P_m$ which have classifications $C_1, C_2, C_3, \ldots, C_m$ respectively. The patterns are represented by the following IFSs in the universe of discourse $X = \{x_1, x_2, \ldots, x_n\}$: $P_i = \{\langle x_j, \mu_{P_i}(x_j), \nu_{P_i}(x_j)\rangle / x_j \in X\}$, where $i = 1, 2, \ldots, m$ and $j = 1, 2, \ldots, n$.

Given an unknown pattern Q, represented by IFS

$$Q_i = \{\langle x_j, \mu_{Q_i}(x_j), \nu_{Q_i}(x_j)\rangle / x_j \in X\}.$$

Our aim here is to classify Q to one of the classes $C_1, C_2, C_3, \ldots, C_m$. According to the principle of minimum divergence degree between IFSs [19], the process of assigning Q to C_{k^*} is described by

$$k^* = \arg\min_k \{D(P_k, Q)\}.$$

According to this algorithm, the given pattern can be recognized so that the best class can be selected.

Example 7.1 Given three known patterns P_1, P_2 and P_3 which have classifications C_1, C_2 and C_3 respectively. These are represented by the following IFSs in the universe of discourse $X = \{x_1, x_2, x_3\}$:

$$P_1 = \{(x_1, 0.2, 0.5), (x_2, 0.5, 0.4), (x_3, 0.2, 0.4)\},$$
$$P_2 = \{(x_1, 0.4, 0.3), (x_2, 0.6, 0.1), (x_3, 0.5, 0.2)\},$$
$$P_3 = \{(x_1, 0.1, 0.4), (x_2, 0.3, 0.5), (x_3, 0.7, 0.1)\},$$

where for $i = 1, 2, 3$

$$P_i = \{\langle x_1, \mu_{P_i}(x_1), \nu_{P_i}(x_1)\rangle, \langle x_2, \mu_{P_i}(x_2), \nu_{P_i}(x_2)\rangle, \langle x_3, \mu_{P_i}(x_3), \nu_{P_i}(x_3)\rangle\}.$$

Table 7.1 Comparison of different intuitionistic fuzzy divergence measures

Q	P_1	P_2	P_3
$D^E_{IFS}(P_i, Q)$	0.1463	0.1006	0.1104
$d(P_i, Q)$	0.5500	0.4000	0.5000
$d_H(P_i, Q)$	0.2667	0.2000	0.2667
$D_{IFS}(P_i, Q)$	0.3839	0.1079	0.2246
$D^*(P_i, Q)$	0.1064	0.0786	0.0888

we have an unknown pattern Q, represented by IFS

$$Q = \{(x_1, 0.4, 0.5), (x_2, 0.3, 0.2), (x_3, 0.6, 0.3)\},$$

our aim here is to classify Q to one of the classes C_1, C_2 and C_3. From the formulae (7.7), (7.12)–(7.15), we compute the values of different intuitionistic fuzzy divergence measures and are presented in Table 7.1.

From the computed numerical values of different existing measures and the proposed measure D^E_{IFS} presented in Table 7.1, it is observed that the pattern Q should be classified to C_2. Thus, the proposed intuitionistic fuzzy exponential divergence measure is consistent for the application point of view in the context of pattern recognition.

7.5 Application of Intuitionistic Fuzzy Exponential Divergence in Multi-attribute Decision-Making

In this section, we present the application of the proposed intuitionistic fuzzy exponential divergence measure in the field of multi-attribute decision-making. It is widely known that the decision-making problem is the process of finding the best option from all of the feasible alternatives. For this let us assume that there exists a set $A = \{A_1, A_2, A_3, \ldots, A_m\}$ of m alternatives and another set of n attributes given by $M = \{M_1, M_2, M_3, \ldots, M_n\}$. The decision-maker has to find the best alternative from the set A corresponding to the set M of n attributes. Further, suppose that $D = (d_{ij})_{n \times m}$ is the intuitionistic fuzzy decision matrix, where $d_{ij} = (\mu_{ij}, \nu_{ij})$ is an attribute provided by the decision-maker.

$$\mu_{ij} = \text{degree for which attribute } M_i \text{ is satisfied by the alternative } A_j,$$
$$\nu_{ij} = \text{degree for which attribute } M_i \text{ is not satisfied by the alternative } A_j.$$

The computational procedure to solve the intuitionistic fuzzy multi-attribute decision-making problem is as follows:

Step I: Construct the Normalized Decision Matrix
This step converts the various dimensional attributes into non-dimensional attributes, as per the method provided by [25]. In general, all of the attributes may be of the same type or different types. If the attributes are of different types, it is required to make them of the same type. For example, it is assumed that there are two types of attributes, say (i) benefit type, and (ii) the cost type. Depending on the nature of attributes we convert the cost attribute into the benefit attribute. So we transform the intuitionistic fuzzy decision matrix $D = (d_{ij})_{n\times m}$ into the normalized intuitionistic fuzzy decision matrix say $R = (r_{ij})_{n\times m}$. An element r_{ij} of the normalized decision matrix R is obtained as follows:

$$r_{ij} = (\mu_{ij}, \nu_{ij}) = \begin{cases} d_{ij}, \ \textit{for benefit attribute}, \ M_i \\ d_{ij}^c, \ \textit{for cos t attribute}, M_i \end{cases}; \tag{7.16}$$

where $i = 1, 2, \ldots, n; \ j = 1, 2, \ldots, m.$
and d_{ij}^c is the complement of d_{ij} with $d_{ij}^c = (\nu_{ij}, \mu_{ij})$.

Step II: Specify the Options by the Characteristic Sets
With the normalized matrix $R = (r_{ij})_{n\times m}$ we specify the option A_j, by the characteristic sets given by

$$A_j = \{\langle M_i, \mu_{ij}, \nu_{ij}\rangle / M_i \in M\}, \quad i = 1, 2, \ldots, n, \ j = 1, 2, \ldots, m \tag{7.17}$$

Step III: Determine the Ideal Solution A^*

$$A^* = \{(\mu_1^*, \nu_1^*), (\mu_2^*, \nu_2^*), (\mu_3^*, \nu_3^*) \ldots, (\mu_n^*, \nu_n^*)\} \tag{7.18}$$

$$\text{where } (\mu_i^*, \nu_i^*) = (\max_j \mu_{ij}, \min_j \nu_{ij}) \quad \text{with } i = 1, 2, \ldots, n. \tag{7.19}$$

Step IV: Calculate the Divergence using $D_{IFS}^E(A, B)$ Given in (7.7)
Calculate the divergence using the following expression:

$$D_{IFS}^E(A_j, A^*) = \sum_{i=1}^{n} \left[\begin{array}{l} 2 - \left(1 - \frac{(\mu_{A_j}(x_i) - \mu_{A^*}(x_i)) - (\nu_{A_j}(x_i) - \nu_{A^*}(x_i))}{2}\right) \\ e^{\frac{(\mu_{A_j}(x_i) - \mu_{A^*}(x_i)) - (\nu_{A_j}(x_i) - \nu_{A^*}(x_i))}{2}} \\ -\left(1 + \frac{(\mu_{A_j}(x_i) - \mu_{A^*}(x_i)) - (\nu_{A_j}(x_i) - \nu_{A^*}(x_i))}{2}\right) \\ e^{\frac{(\nu_{A_j}(x_i) - \nu_{A^*}(x_i)) - (\mu_{A_j}(x_i) - \mu_{A^*}(x_i))}{2}} \end{array} \right], \tag{7.20}$$

$\forall j = 1, 2, \ldots m$

Table 7.2 Intuitionistic fuzzy decision matrix D

	A_1	A_2	A_3	A_4	A_5
M_1	(0.4, 0.5)	(0.7, 0.2)	(0.6, 0.1)	(0.5, 0.4)	(0.4, 0.3)
M_2	(0.8, 0.1)	(0.5, 0.3)	(0.7, 0.3)	(0.3, 0.4)	(0.7, 0.1)
M_3	(0.7, 0.3)	(0.3, 0.4)	(0.6, 0.2)	(0.8, 0.1)	(0.5, 0.2)
M_4	(0.6, 0.2)	(0.8, 0.1)	(0.4, 0.1)	(0.7, 0.2)	(0.9, 0.1)
M_5	(0.5, 0.4)	(0.2, 0.6)	(0.3, 0.4)	(0.6, 0.1)	(0.8, 0.0)
M_6	(0.3, 0.4)	(0.4, 0.5)	(0.8, 0.2)	(0.7, 0.1)	(0.6, 0.4)

Step V: Rank the Order of Preference
The best alternative A_k is now obtained corresponding to the smallest value of the divergence measure $D^E_{IFS}(A_j, A^*)$, $\forall j = 1, 2, \ldots m$.

Now the application of introducing intuitionistic fuzzy exponential divergence measure $D^E_{IFS}(A, B)$ is demonstrated with the help of a numerical example below

Example 7.2 Let us assume that a car company wants to select a suitable material supplier among five alternative suppliers A_1, A_2, A_3, A_4 and A_5 while considering six attributes: (i) Quality of Product (M_1), (ii) Price (M_2), (iii) Technical Capability (M_3), (iv) Delivery (M_4), (v) Service (M_5), and (vi) Flexibility (M_6).

Table 7.2 shows the intuitionistic fuzzy decision matrix $D = (d_{ij})_{6\times5}$ having the characteristics of the alternatives $A_j(j = 1, 2, 3, 4, 5)$.

Step I: Here M_2 is the cost attribute while other five are benefit attributes. We transform the cost attribute M_2 into benefit attribute M_2^c we get

$$M_2^c = \{(0.1,\ 0.8), (0.3,\ 0.5), (0.3,\ 0.7), (0.4,\ 0.3), (0.1,\ 0.7)\}$$

Table 7.3 presents the normalized intuitionistic fuzzy decision matrix R.
Step II: Characteristic sets presenting the options A_j are given by

$$A_1 = \{(M_1, 0.4, 0.5), (M_2, 0.1, 0.8), (M_3, 0.7, 0.3), (M_4, 0.6, 0.2), (M_5, 0.5, 0.4), (M_6, 0.3, 0.4)\},$$
$$A_2 = \{(M_1, 0.7, 0.2), (M_2, 0.3, 0.5), (M_3, 0.3, 0.4), (M_4, 0.8, 0.1), (M_5, 0.2, 0.6), (M_6, 0.4, 0.5)\},$$
$$A_3 = \{(M_1, 0.6, 0.1), (M_2, 0.3, 0.7), (M_3, 0.6, 0.2), (M_4, 0.4, 0.1), (M_5, 0.3, 0.4), (M_6, 0.8, 0.2)\},$$
$$A_4 = \{(M_1, 0.5, 0.4), (M_2, 0.4, 0.3), (M_3, 0.8, 0.1), (M_4, 0.7, 0.2), (M_5, 0.6, 0.1), (M_6, 0.7, 0.1)\},$$
$$A_5 = \{(M_1, 0.4, 0.3), (M_2, 0.1, 0.7), (M_3, 0.5, 0.2), (M_4, 0.9, 0.1), (M_5, 0.8, 0.0), (M_6, 0.6, 0.4)\}.$$

Table 7.3 Normalized intuitionistic fuzzy decision matrix R

	A_1	A_2	A_3	A_4	A_5
M_1	(0.4, 0.5)	(0.7, 0.2)	(0.6, 0.1)	(0.5, 0.4)	(0.4, 0.3)
M_2^c	(0.1, 0.8)	(0.3, 0.5)	(0.3, 0.7)	(0.4, 0.3)	(0.1, 0.7)
M_3	(0.7, 0.3)	(0.3, 0.4)	(0.6, 0.2)	(0.8, 0.1)	(0.5, 0.2)
M_4	(0.6, 0.2)	(0.8, 0.1)	(0.4, 0.1)	(0.7, 0.2)	(0.9, 0.1)
M_5	(0.5, 0.4)	(0.2, 0.6)	(0.3, 0.4)	(0.6, 0.1)	(0.8, 0.0)
M_6	(0.3, 0.4)	(0.4, 0.5)	(0.8, 0.2)	(0.7, 0.1)	(0.6, 0.4)

Table 7.4 Values of $D^E_{IFS}(A_j, A^*)$ for $A_j (j = 1, 2, 3, 4, 5)$

$D^E_{IFS}(A_1, A^*)$	$D^E_{IFS}(A_2, A^*)$	$D^E_{IFS}(A_3, A^*)$	$D^E_{IFS}(A_4, A^*)$	$D^E_{IFS}(A_5, A^*)$
1.2830	1.4764	0.7266	**0.2217**	0.5881

Step III: The ideal solution A^* obtained using Eqs. (7.18) and (7.19) is given by

$$A^* = \{(0.7, 0.1), (0.4, 0.3), (0.8, 0.1), (0.9, 0.1), (0.8, 0.0), (0.8, 0.1)\}$$

Step IV: Table 7.4 gives the values of $D^E_{IFS}(A_k, A^*)$ using measure (7.20).
Step V: The best alternative is A_4 obtained corresponding to the smallest value of divergence measure $D^E_{IFS}(A_j, A^*)$ for $j = 4$.

Hence, the numerical example shows that the proposed intuitionistic fuzzy exponential divergence measure is a very suitable measure to solve the multi-attribute decision-making problems.

7.6 Application of Intuitionistic Fuzzy TOPSIS and MOORA Methods for Multi-attribute Decision-Making: A Comparative Analysis

We now present the application of proposed intuitionistic fuzzy exponential divergence measure in context of multi-attribute decision-making using TOPSIS [13] and MOORA [5, 20] methods in intuitionistic fuzzy environment.

7.6.1 Intuitionistic Fuzzy TOPSIS Method

Let $A = \{A_1, A_2, A_3, \ldots, A_m\}$ be a set of m alternatives and decision-maker will choose the best one from A according to an attribute set given by $M = \{M_1, M_2, M_3, \ldots, M_n\}$.

Various computational steps in the intuitionistic fuzzy TOPSIS method are as follows:

1. Construction of intuitionistic fuzzy decision matrix
   In this step, intuitionistic fuzzy decision matrix $D = (d_{ij})_{n \times m}$ of intuitionistic fuzzy value $d_{ij} = (\mu_{ij}, \nu_{ij})$ is constructed.
2. Construct the normalized intuitionistic fuzzy decision matrix $R = (r_{ij})_{n \times m}$
   The normalized value r_{ij} is calculated as

$$r_{ij} = (\mu_{ij}, v_{ij}) = \left(\mu_{ij} \Big/ \sqrt{\sum_{j=1}^{m} \mu_{ij}^2}, v_{ij} \Big/ \sqrt{\sum_{j=1}^{m} v_{ij}^2} \right), \qquad (7.21)$$

$$i = 1, 2, \ldots, n, \; j = 1, 2, \ldots, m$$

3. Construct the weighted normalized intuitionistic fuzzy decision matrix
   The weighted normalized value

$$v_{ij} = w_i r_{ij}, \quad i = 1, 2, \ldots, n, \; j = 1, 2, \ldots, m \qquad (7.22)$$

   where the weight matrix for each attribute is as follows: $W = [1, 1, 1, 1, 1]$ and w_i is weight or preference value of ith attribute.
4. Determine the intuitionistic fuzzy positive-ideal solution (IFPIS) and intuitionistic fuzzy negative ideal solution (IFNIS)

$$\text{IFPIS} = A^+ = \{v_1^+, v_2^+, \ldots, v_n^+\} = \{(\max_j v_{ij}/i \in J), (\min_j v_{ij}/i \in J')\} \qquad (7.23)$$

$$\text{IFNIS} = A^- = \{v_1^-, v_2^-, \ldots, v_n^-\} = \{(\min_j v_{ij}/i \in J), (\max_j v_{ij}/i \in J')\} \qquad (7.24)$$

   where J is associated with the benefit attribute and J' is associated with the cost attribute.
5. Calculate the separation measures D_{IFS}^{E+} and D_{IFS}^{E-} using the divergence measure in (7.7).
6. Calculate the relative closeness of the ideal solution.
   The relative closeness of alternative A_j with respect to PIS is defined by

$$C_j = \frac{D_{IFS}^{E-}}{D_{IFS}^{E-} + D_{IFS}^{E+}}, \quad j = 1, 2, \ldots, m. \qquad (7.25)$$

7. Rank the preference order of all alternatives according to the closeness coefficient.

Now the application of proposed measure $D_{IFS}^{E}(A, B)$ with TOPSIS technique is demonstrated using the intuitionistic fuzzy decision matrix considered in Table 7.2.

Table 7.5 presents the normalized/weighted intuitionistic fuzzy decision matrix corresponding to the intuitionistic fuzzy decision matrix given in Table 7.2 using the formulae (7.21) and (7.22).

Table 7.6 shows the intuitionistic fuzzy positive and negative ideal solutions A^+ and A^- using formulae (7.23) and (7.24).

Table 7.5 Normalized/weighted intuitionistic fuzzy decision matrix R

	A_1	A_2	A_3	A_4	A_5
M_1	(0.2434, 0.3043)	(0.4358, 0.1245)	(0.3833, 0.0639)	(0.3037, 0.2430)	(0.2302, 0.1726)
M_2	(0.4868, 0.0609)	(0.3113, 0.1868)	(0.4472, 0.1917)	(0.1822, 0.2430)	(0.4028, 0.0575)
M_3	(0.4259, 0.1826)	(0.1868, 0.2490)	(0.3833, 0.1278)	(0.4860, 0.0607)	(0.2877, 0.1151)
M_4	(0.3651, 0.1217)	(0.4981, 0.0622)	(0.2556, 0.0639)	(0.4252, 0.1215)	(0.5179, 0.0575)
M_5	(0.3043, 0.2434)	(0.1245, 0.3735)	(0.1917, 0.2556)	(0.3645, 0.0607)	(0.4603, 0.000)
M_6	(0.1826, 0.2434)	(0.2490, 0.3113)	(0.5111, 0.1278)	(0.4252, 0.0607)	(0.3453, 0.2302)

Table 7.6 Fuzzy positive and negative ideal solutions

	A^+	A^-
M_1	(0.4358, 0.0639)	(0.2302, 0.3043)
M_2	(0.4868, 0.0575)	(0.1822, 0.2430)
M_3	(0.4860, 0.0607)	(0.1868, 0.2490)
M_4	(0.5179, 0.0575)	(0.2556, 0.1217)
M_5	(0.4603, 0.0000)	(0.1245, 0.3735)
M_6	(0.5111, 0.0607)	(0.1826, 0.3113)

The calculated numerical values of separation measures of each alternative from positive-ideal solution and negative solution using measure (7.7) are given in Table 7.7.

The calculated values of relative closeness of each alternative to positive-ideal solution using the formula (7.25) and their corresponding ranks are presented in Table 7.8.

According to the closeness coefficient and ranking of alternative, it is obtained that A_5 is the best alternative.

Table 7.7 Numerical values of separation measures D_{IFS}^{E+} and D_{IFS}^{E}

	D_{IFS}^{E+}	D_{IFS}^{E-}
A_1	0.3533	0.2278
A_2	0.5834	0.1419
A_3	0.2137	0.3374
A_4	0.2025	0.4345
A_5	0.1433	0.4787

Table 7.8 Closeness coefficient and ranking of alternatives

	C_j	Rank
A_1	0.3920	4
A_2	0.1419	5
A_3	0.3374	3
A_4	0.4345	2
A_5	**0.4787**	**1**

7.6.2 Intuitionistic Fuzzy MOORA Method

Intuitionistic fuzzy MOORA method for solving multi-attribute decision-making is as follows. The computational procedure in intuitionistic fuzzy MOORA method up to step 3 is same as discussed in TOPSIS method above.

Step 4: Compute the overall rating M^+ and M^- of benefit and cost attribute for each alternative from Table 7.5.

$$M^+ = \{v_1^+, v_2^+, \ldots, v_n^+\} = \{(\max_i v_{ij}, \min_i v_{ij})/j \in J\} \tag{7.26}$$

$$M^- = \{v_1^-, v_2^-, \ldots, v_n^-\} = \{(\min_i v_{ij}, \max_i v_{ij})/j \in J\}, \tag{7.27}$$

where $J = \{j = 1, 2, 3, \ldots, m/j$ is associated with the alternative$\}$.

The calculated overall rating M^+ and M^- of each of alternative are shown in Table 7.9.

Step 5: Calculate the overall performance index $D^E_{IFS}(M^+, M^-)$ for each alternative using the formula (7.7) and the computed values are given in Table 7.9.

Step 6: Ranking alternatives and/or selecting the most efficient one based on the value of $D^E_{IFS}(M^+, M^-)$.

Table 7.10 presents the overall performance index $D^E_{IFS}(M^+, M^-)$ for each alternatives and their ranking.

According to the calculated results and ranking order of the alternatives, it is obtained that A_4 is the most preferable alternative.

Table 7.9 Overall rating M^+ and M^- of each alternative

	M^+	M^-
A_1	(0.4868, 0.0609)	(0.1826, 0.3043)
A_2	(0.4981, 0.0622)	(0.1245, 0.3735)
A_3	(0.5111, 0.0639)	(0.1917, 0.2556)
A_4	(0.4860, 0.0607)	(0.1822, 0.2430)
A_5	(0.5179, 0.0000)	(0.2302, 0.2302)

Table 7.10 Closeness coefficient and ranking

	$D^E_{IFS}(M^+, M^-)$	Rank
A_1	0.1575	4
A_2	0.2533	5
A_3	0.1364	2
A_4	**0.1229**	**1**
A_5	0.1402	3

7.6.3 A Comparative Analysis of the Proposed Method and the Existing Methods of Multi-attribute Decision-Making

We now compare the proposed method of multi-attribute decision-making with the existing methods of multi-attribute decision-making using the proposed measure (7.7). From the proposed method in Sect. 7.6, it is obtained that the alternative A_4 is more preferable alternative for a car company among five. However, we have examined from the TOPSIS method that A_5 is more preferable alternative and from MOORA method we have obtained that A_4 is a more preferable alternative which is exactly same as the result given by our new method. It is shown that the proposed method is very simple, consistent and efficient among the compared methods.

7.7 Concluding Remarks

In this chapter, we have proposed an information-theoretic exponential framework for IFSs and established some of the properties of the proposed intuitionistic fuzzy exponential divergence measure. It is shown that the proposed divergence measure is feasible and efficient in point of view of pattern recognition. Moreover, a method is developed to solve MADM problems based on the proposed divergence measure under intuitionistic fuzzy environment. In addition, the application of the proposed divergence measure is presented in two existing TOPSIS and MOORA methods of MADM in an intuitionistic fuzzy environment. The consistency of the results of the proposed method is shown by a comparative study of the proposed method and the existing methods of MADM. Finally, we note that the proposed divergence measure is a very appropriate measure to solve the real-world problems related to MADM.

References

1. Atanassov KT (1986) Intuitionistic fuzzy sets. Fuzzy Sets Syst 20:87–96
2. Atanassov KT (1994) New operations defined over the intuitionistic fuzzy sets. Fuzzy Sets Syst 61:137–142
3. Atanassov KT (1999) Intuitionistic fuzzy sets. Springer, Heidelberg
4. Atanassov KT (2000) Two theorems for intuitionistic fuzzy sets. Fuzzy Sets Syst 110: 267–269
5. Brauers WKM, Zavadskas EK (2006) The MOORA method and its application to privatization in transition economy. Control Cybern 35(2):443–468
6. Bustince H, Burillo P (1996) Vague sets are intuitionisic fuzzy sets. Fuzzy Sets Syst 79: 403–405
7. De SK, Biswas R, Roy AR (2001) An application of intuitionistic fuzzy sets in medical diagnosis. Fuzzy Sets Syst 117(2):209–213

8. Fan J, Xie W (1999) Distance measures and induced fuzzy entropy. Fuzzy Sets Syst 104 (2):305–314
9. Gau WL, Buehrer DJ (1993) Vague sets. IEEE Trans Syst Man Cybern 23:610–614
10. Hatzimichailidis AG, Papakostas GA, Kaburlasos VG (2012) A novel distance measure of intuitionistic fuzzy sets and its application to pattern recognition problems. Int J Intell Syst 27(4):396–409
11. Hung WL, Yang MS (2004) Similarity measures of intuitionistic fuzzy sets based on Hausdorff distance. Pattern Recogn Lett 25:1603–1611
12. Hung WL, Yang MS (2008) On the J-divergence of intuitionistic fuzzy sets and its application to pattern recognition. Inf Sci 178(6):1641–1650
13. Hwang CL, Yoon K (1981) Multiple attribute decision making–methods and applications. Springer, New York
14. Jiang YC, Tang Y, Wang J, Tang S (2009) Reasoning within intuitionistic fuzzy rough description logics. Inf Sci 179:2362–2378
15. Li F, Lu ZH, Cai LJ (2003) The entropy of vague sets based on fuzzy sets. J Huazhong Univ Sci Technol (Nature Science) 31:1–3
16. Li DF (2004) Some measures of dissimilarity in intuitionistic fuzzy structures. J Comput Syst Sci 68(1):115–122
17. Li DF (2005) Multi-attribute decision-making models and methods using intuitionistic fuzzy sets. J Comput Syst Sci 70(1):73–85
18. Papakostas GA, Hatzimichailidis AG, Kaburlasos VG (2013) Distance and similarity measures between intuitionistic fuzzy sets: a comparative analysis from a pattern recognition point of view. Pattern Recogn Lett 34(14):1609–1622
19. Shore JE, Gray RM (1982) Minimization cross-entropy pattern classification and cluster analysis. IEEE Trans Pattern Anal Mach Intell 4(1):11–17
20. Stanujkic D, Magdalinovic N, Tojanovic S, Jovanovic R (2012) Extension of ratio system part of MOORA method for solving decision making problems with interval data. Informatica 23(1):141–154
21. Verma R, Sharma BD (2012) On generalized intuitionistic fuzzy divergence (relative information) and their properties. J Uncertain Syst 6(4):308–320
22. Verma R, Sharma BD (2013) Intuitionistic fuzzy jensen-rényi divergence: applications to multiple-attribute decision making. Informatica 37(4):399–409
23. Verma R, Sharma BD (2013) Exponential entropy on intuitionistic fuzzy sets. Kybernetika 49:114–127
24. Vlachos IK, Sergiadis GD (2007) Intuitionistic fuzzy information-application to pattern recognition. Pattern Recogn Lett 28(2):197–206
25. Xu Z, Hu H (2010) Projection models for intuitionistic fuzzy multiple attribute decision making. Int J Inf Technol Decis Making 9(2):267–280
26. Zadeh LA (1965) Fuzzy sets. Inf Control 8:338–353
27. Zhang QS, Jiang SY (2008) A note on information entropy measures for vague sets and its applications. Inf Sci 178:4184–4191
28. Zhang SF, Liu SY (2011) A GRA-based intuitionistic multi-criteria decision making method for personnel selection. Expert Syst Appl 38(9):11401–11405

Epilogue

We hope the reader has enjoyed the material provided in this book. The book focused on generalization of fuzzy information measures, fuzzy divergence measures and intuitionistic fuzzy divergence measures. The characterization and generalization of various measures of fuzzy information were used to introduce and validate new generalized R-norm fuzzy information and divergence measures. The relations were established between generalized fuzzy entropy measure and their fuzzy divergence measures. A new parametric generalized exponential fuzzy divergence measure was obtained in the process of generalizations.

In addition, a comparative application was provided in the context of strategic decision-making. A sequence of fuzzy mean difference divergence measures was introduced with a number of inequalities among them. In this way, their efficiency in pattern recognition and compound linguistic variables was achieved. Moreover, a newly generalized fuzzy divergence measure and its application to multi-criteria decision-making and pattern recognition were provided. Finally, the book addressed the issue of intuitionistic fuzzy set theory—a generalization of fuzzy set theory—with a new definition for exponential divergence measure in intuitionistic fuzzy nature. A number of interesting and elegant properties were found that enhanced the importance of proposed exponential divergence measure. A comparison of the proposed method of multi-attribute decision-making in an intuitionistic fuzzy environment with the existing TOPSIS and MOORA methods was presented with discussion. On an application level, the proposed generalizations of fuzzy information measures open the way to numerous applied areas. It was shown that the proposed measure and method were more appropriate to solve the real-world problems related to multi-attribute decision-making.

The present work is based on both the quantitative and the qualitative generalized fuzzy information measures, divergence measures and intuitionistic fuzzy divergence measures. It is found that these measures are more flexible from application point of view in different fields. Some of the proposed measures could be further compared with existing measures of divergence. The corresponding

A. Ohlan and R. Ohlan, *Generalizations of Fuzzy Information Measures*,
DOI 10.1007/978-3-319-45928-8

measures of similarity in fuzzy and intuitionistic fuzzy environment may be tried with their application in different fields. The method of pattern recognition that is used in fuzzy and intuitionistic fuzzy environment using the generalized divergence measures can be extended to other problems like handwritten character recognition, fingerprint recognition, human face recognition and classification of X-ray images.

Index

A. Ohlan and R. Ohlan, *Generalizations of Fuzzy Information Measures*,
DOI 10.1007/978-3-319-45928-8

Printed in the United States
By Bookmasters